"My ShopBot has completely transformed the way I work. I've gone from being creatively frustrated to being able to reproduce nearly anything using a ShopBot. It's a great feeling to be able to make just about anything you can dream up."
Brady Watson, iBILD, LLC
www.ibild.com
www.shopbottools.com/Brady's Tricks.htm

Six years ago, Brady Watson, owner of iBILD, LLC, left the rat race as a software engineer and charted a new course with this first ShopBot. Since then, he has used his ShopBot to cut nearly every material - from aluminum and carbon fiber composites to high-end 3D architectural millwork - for customers around the world. Brady uses his ShopBot to proof the data he collects on his 3D laser scanners and to create unique, custom made parts for a variety of projects. When he's not "working," he is coming up with innovative ways to assist in the CNC process and contributing regularly to discussions on the TalkShopBot Forum, sharing his knowledge with others.
Brady is a Maker and a teacher.
An artist and an inventor.
Brady is a ShopBotter

What would you like to make today?

ShopBot's new PR⌐alpha BT32 "Buddy."

www.ShopBotTools.com
888-680-4466

Make:

Volume 12

technology on your time ™

ON THE COVER: MAKE engineering intern Matthew Dalton wrangled the Blubber Bots near San Francisco's Bay Bridge, while photographer Doug Adesko captured the autonomous blimps in flight.

Columns

e the clip:

tely!

Vol. 12, Nov. 2007. MAKE (ISSN 1556-2336) is published quarterly by O'Reilly Media, Inc. in the months of March, May, August, and November. O'Reilly Media is located at 1005 Gravenstein Hwy. North, Sebastopol, CA 95472, (707) 827-7000. SUBSCRIP-TIONS: Send all subscription requests to MAKE, P.O. Box 17046, North Hollywood, CA 91615-9588 or subscribe online at makezine.com/offer or via phone at (866) 289-8847 (U.S. and Canada); all other countries call (818) 487-2037. Subscriptions are available for $34.95 for 1 year (4 quarterly issues) in the United States; in Canada: $39.95 USD; all other countries: $49.95 USD. Periodicals Postage Paid at Sebastopol, CA, and at additional mailing offices. POSTMASTER: Send address changes to MAKE, P.O. Box 17046, North Hollywood, CA 91615-9588. Canada Post Publications Mail Agreement Number 41129568. CANADA POSTMASTER: Send address changes to: O'Reilly Media, PO Box 456, Niagara Falls, ON L2E 6V2

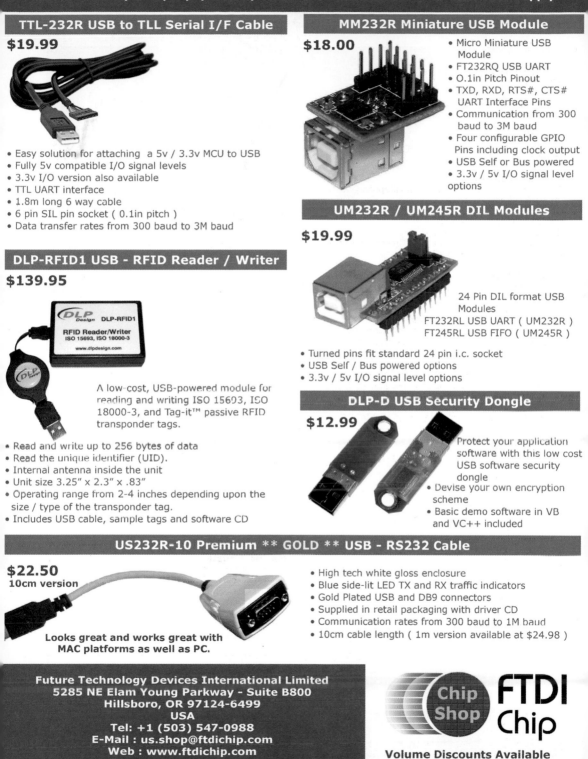

Make: Projects

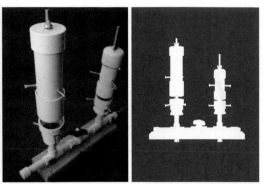
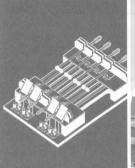

Make:
technology on your time ™
Volume 12

‾ ‾ ‾ ‾ ‾ ‾ ‾ ‾ ‾ ‾ ‾ ‾ ‾ ‾ ‾ ‾ ‾
READ ME: Always check the URL associated
with a project before you get started. There
may be important updates or corrections.

Maker

DIY

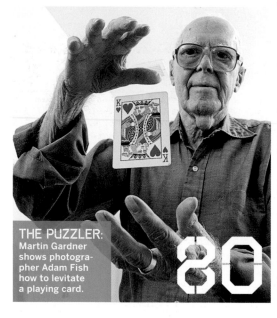

THE PUZZLER:
Martin Gardner
shows photogra-
pher Adam Fish
how to levitate
a playing card.

Make:
technology on your time™

EDITOR AND PUBLISHER
Dale Dougherty
dale@oreilly.com

EDITOR-IN-CHIEF
Mark Frauenfelder
markf@oreilly.com

CREATIVE DIRECTOR
Daniel Carter
dcarter@oreilly.com

MANAGING EDITOR
Shawn Connally
shawn@oreilly.com

DESIGNER
Katie Wilson

ASSOCIATE MANAGING EDITOR
Goli Mohammadi

PRODUCTION DESIGNER
Gerry Arrington

SENIOR EDITOR
Phillip Torrone
pt@makezine.com

PHOTO EDITOR
Sam Murphy
smurphy@oreilly.com

PROJECTS EDITOR
Paul Spinrad
pspinrad@makezine.com

ONLINE MANAGER
Terrie Miller

SECTION EDITOR
Charles Platt

ASSOCIATE PUBLISHER
Dan Woods
dan@oreilly.com

STAFF EDITOR
Arwen O'Reilly

CIRCULATION DIRECTOR
Heather Harmon

COPY CHIEF
Keith Hammond

ACCOUNT MANAGER
Katie Dougherty

EDITOR AT LARGE
David Pescovitz

MARKETING & EVENTS COORDINATOR
Rob Bullington

VIDEO PRODUCER
Bre Pettis

MAKE TECHNICAL ADVISORY BOARD
Evil Mad Scientist Laboratories, Limor Fried, Joe Grand, Saul Griffith, Bunnie Huang, Tom Igoe, Steve Lodefink, Erica Sadun

PUBLISHED BY O'REILLY MEDIA, INC.
Tim O'Reilly, CEO
Laura Baldwin, COO

Visit us online at makezine.com
Comments may be sent to editor@makezine.com

For advertising inquiries, contact:
Katie Dougherty, 707-827-7272, katie@oreilly.com

For sponsorship inquiries, contact:
Scott Feen, 707-827-7105, scottf@oreilly.com

For event inquiries, contact:
Sherry Huss, 707-827-7074, sherry@oreilly.com

PLEASE NOTE: Technology, the laws, and limitations imposed by manufacturers and content owners are constantly changing. Thus, some of the projects described may not work, may be inconsistent with current laws or user agreements, or may damage or adversely affect some equipment. Your safety is your own responsibility, including proper use of equipment and safety gear, and determining whether you have adequate skill and experience. Power tools, electricity, and other resources used for these projects are dangerous, unless used properly and with adequate precautions, including safety gear. Some illustrative photos do not depict safety precautions or equipment, in order to show the project steps more clearly. These projects are not intended for use by children.
Use of the instructions and suggestions in MAKE is at your own risk. O'Reilly Media, Inc., disclaims all responsibility for any resulting damage, injury, or expense. It is your responsibility to make sure that your activities comply with applicable laws, including copyright.

Contributing Editors: Gareth Branwyn, William Gurstelle, Mister Jalopy, Brian Jepson

Contributing Artists: Doug Adesko, Scott Barry, Kay Canavino, Roy Doty, Nick Dragotta, Julian Feinberg, Adam Fish, Dustin Amery Hostetler, Timmy Kucynda, Tim Lillis, Jason Madara, Charles Platt, Nik Schulz, Jen Siska, Jonathan Sprague, Robert Stegmann

Contributing Writers: John Alderman, Mark Allen, Chris Anderson, Tim Anderson, T.J. "Skip" Arey, Bill Barminski, Joost Bonsen, Ed Bringas, Annie Buckley, Michael A. Covington, Kindy Connally-Stewart, Jérôme Demers, Cory Doctorow, George Dyson, Lenore Edman, Dhananjay V. Gadre, Melissa Gira, Saul Griffith, Michelle Hlubinka, Martin Howse, Daryl Hrdlicka, Parker Jardine, Richard Kadrey, Kevin Kelly, Katie Kurtz, Norene Leddy, Tim Lillis, Andrew Milmoe, Dave Ng, Rory Nugent, Brian O'Heir, Windell Oskay, Tom Owad, John Edgar Park, Tom Parker, José Pino, Michael H. Pryor, Douglas Repetto, Rudy Rucker, Donald E. Simanek, David Simpson, Bruce Sterling, Bruce Stewart, Jason Torchinsky, Cy Tymony, Megan Mansell Williams, Lee Zlotoff

Interns: Matthew Dalton (engr.), Adrienne Foreman (web), Arseny Lebedev (web)

Customer Service cs@readerservices.makezine.com
Manage your account online, including change of address at: makezine.com/account
866-289-8847 toll-free in U.S. and Canada
818-487-2037, 5 a.m.–5 p.m., PST

Contributors

Rudy Rucker (*Cellular Automata*) has worked as a mathematics professor, a software engineer, a computer science professor, and a freelance writer. He's published 29 books, including a nonfiction book on the meaning of computers: *The Lifebox, the Seashell and the Soul*. He publishes an online SF zine called *Flurb*, and has been known to say that everything is a cellular automaton. He's currently writing a cyberpunkish trilogy of novels in which nanotechnology changes everything. The first, *Postsingular*, appeared from Tor this fall, and will soon be available for free download on the web. rudyrucker.com

"Despite being the editor and publisher of Cool Tools," **Kevin Kelly** (*Book Yourself*) admits, "I'm neo-Amish. I am selectively late in adopting new technology, and mindful of the social costs when I do. I like old tools and gadgets that sport wear-marks and other evidence they are really used. Rule of thumb: never put a cover case around anything handy. That includes iPods. And when it comes to shopping, thrift counts. 'High quality' is usually overrated. With rare exceptions, if Costco sells it, it is good enough for me."

Parker Jardine (*Primer*) says, "Picture a super-strong, blond-haired brute that enjoys every sport the Southwest has to offer. Then couple that with a 7 to 5 weekday job as a systems administrator. I spend most of my spare time either working on my renewable energy projects, or on epic, multiday mountain bike rides with my girlfriend, Jennifer, and super dog, Lilly." He's currently working on connecting his solar PV system to the utility grid and designing a solar panel for electrolysis, and he is fond of top sirloin elk steak.

Richard Kadrey (*Infrared Photography*) is a freelance writer living in San Francisco. He has written about art, culture, and technology for places such as *Wired*, Discovery Online, and *Wired for Sex* on the G4 cable network. He is the author of the new novel *Butcher Bird*, as well as three other novels, including the cult favorite *Metrophage*. Along with the Pander Brothers, he developed the original comic *Accelerate* for DC/Vertigo. Image Comics will reprint the comic as a graphic novel in fall 2007. Kadrey also works as a photographer under the name KaosBeautyKlinik. He is working on a new novel.

Rory Nugent (*Solar Xylophone*) began creating electronic art after starting NYU's ITP program over a year ago. Previously "stuck in a remote factory somewhere in backwoods North Carolina," he's now completely immersed in schoolwork. These days, "between forgetting to eat lunch and getting no sleep, I obsess over solar panels, wind turbines, and growing tiny potted plants, wishing I had the time in my day to play the Wii, my massive Netflix queue, and 'so bad it's good' dance music." He likes thinking about "all the things it's easy to forget about when our lives are moving at 100mph" and "passé technology like FM radio and analog telephones."

Adam Fish (*Mathemagician* photography) got his start in photography "by despising the graphic design program at UNT," where, as a typography student, he resented having to spend "hours and hours inking letters with Rapidographs." He currently lives in Dallas with his wife, Brooke, and son, Jude, and is a fan of sports, both watching and playing. When asked what new idea excites him most in or out of his field, he says that "although not new, taking a nap sounds pretty exciting to me. It also seems to be outside the field of parenting." He recently shot a portrait of George Foreman in a marching band costume.

You can't learn everything from a book.

In the web-based O'Reilly School of Technology, you learn by doing. In our courses and accredited certificate programs, instructors provide feedback and encouragement, while students experiment with emerging technologies and build real-world projects.

Get the skills and experience you need to succeed. Find out about the OST today at **http://oreillyschool.com**. Use code **ORALL1** for a **30% group member discount**.

Welcome

SPARE PARTS ARE ESSENTIAL

By Mark Frauenfelder

CLUTTER IS A DEMOTIVATING ENERGY sapper. Every couple of months, when my desk gets stacked high with papers, gadgets, and periodicals, and my office floor has accumulated piles of books and packages, I conduct an area sweep collecting the things I don't have an obvious need for. Mostly I throw them in the trash or donate them.

But I don't get rid of everything. Some of the things I come across seem to plead for a second chance at being useful: "Save me! One day you'll be glad you did." These go into a plastic bin I keep in my storage shed.

What kind of things do I put in my spare parts bin? Anything that seems like it might come in handy one day: toys with electric motors, little speakers, switches, Altoids tins, those plastic bubbles used in vending machines to hold little trinkets, broken flashlights, and extra parts from previous projects. The box is a cluttered mess, but it's a contained clutter, and it's actually inspirational. When I poke around in it I dream of possibilities.

These components have come in handy on several occasions. When I made the Vibrobot (see *MAKE, Volume 10, page 119*) I had all the materials I needed on hand. If the project had required a trip to the hardware or electronics store, I might have never completed it.

Same for the Boing Box (*this volume, page 116*), a sound effects prop from the 1951 book *Radio and Television Sound Effects* by Robert B. Turnbull. Because I already had everything I needed, from a wooden cigar box to a spool of galvanized wire, I was able to whip it together in under an hour, and was happily plucking boinging sounds for the remainder of the afternoon.

In both instances I had to modify the project because the stuff in the parts bin didn't quite match the idea I had in my mind, or the plans as printed, but I believe the things I made were better, not worse, because of it.

It's been only a matter of months since I started appreciating the benefits of keeping a bin of spare parts, but master makers have long known how essential it is to the creative process. When I visit their workshops, I've noticed their stockpiles of

stuff with no immediately apparent purpose: parts, scraps, and retired gadgets just waiting for the day when their owner comes up with the idea that calls them into service.

As you look through the projects in this issue of MAKE, think about how you might be able to build them using the stuff you already have lying around. After all, that's how MAKE's authors create a lot of their projects. They use the materials they have at hand.

By improvising, you'll not only avoid a trip to the hardware store, you'll end up making something more personal and possibly better than if you had followed the instructions to the letter.

Take a photo of your parts bin, and/or your final creation, and post a link to it on our comments board at makezine.com/12/welcome. And feel free to add your photos to the MAKE Flickr pool.

Mark Frauenfelder is editor-in-chief of MAKE.

Photograph by Mark Frauenfelder

KNOEND DESIGN...

Hailing from the city by the bay, Knoend Design is a collective of uber-talented folks who create beautiful furniture that pleases both the modernists and Mother Nature. Take the Lite2Go, for example. A totally self-sufficient piece, it arrives in special packaging that converts into a lampshade in about 30 seconds. So which came first, the lamp or the shade?

SITS DOWN WITH CARLO ROSSI.

We gave Knoend a few empty wine jugs and they gave us the Solar Jug Bench. Not bad. Utilizing two empty 4L jugs, recycled truck bed liner and solar powered LED lighting, this bench is an eco-wonder. And it's a great place to kick back once the sun goes down and the jugs light up. But you don't have to be a master craftsman to make something from a jug. All that's required is a few empty bottles and a lot of imagination. For more jug creations and great wine info visit **www.carlorossi.com**. Now go Make Something!

Carlo Rossi

The Instruments of Invention

Bob Dylan was born in his hometown, but Duluth performance artist **Tim Kaiser** has a different musical hero: Harry Partch (1901–1974), an underappreciated composer who invented new microtonal scales for instruments he built himself.

"He was a curmudgeon and a brilliant musician who couldn't stand convention and created his own," says Kaiser, who also coaxes foreign sounds from far-fetched equipment made by hand.

As a teenage musician, Kaiser discovered a new auditory universe at the University of Minnesota and began assembling avant-garde noisemakers to suit his sonic tastes. His technique? Scrap parts and a junior high school electronics class.

Some 20 years later, Kaiser has made more than 150 instruments, including a stenography keyboard wired with the guts of a mini teaching piano, a green effect box with beehive lenses that loops a 2-second delay, and an old espresso bin called *TankPodDrum*, fitted with all things pluckable and tappable. Kaiser takes commissions, but saves his favorites for his own live shows.

TankPodDrum's shell is a hollow, 6"-diameter, 14"-tall stainless steel vessel that Kaiser scored for 70 cents at a salvage yard. In his home studio, he used stove bolts to add a right angle fitting from a hot water heater, brass bells from a rotary phone, a comb of rods from a toy piano, music box tines, bits of chrome, and rack handles. When Kaiser bangs on the attachments with a mallet, the drum acts as a resonator. A pickup epoxied to the barrel's interior connects to an amp or, if Kaiser is playing, a modulation delay that echoes and fades not only the pitch but also the frequency.

After Partch died, the American Composers Forum inherited the rights to his work and released more than 100 of his recordings on the Innova record label. "I've always dreamed of being on Innova," Kaiser says.

Dreams apparently come true. In June 2007, Kaiser's latest solo album, *Analog*, was released on — you guessed it — Innova.

—Megan Mansell Williams

➕ **Watch and listen to Tim Kaiser:** timkaiser.org

Sustainable Junkyard Wars

If you happen to live in rural Bolivia, building a water pump isn't going to include a visit to the local hardware store or being able to plug into a power grid to operate a machine. So how do you get water?

This was the challenge given to **Kara Serenius**, **Hessam Khajeei**, **Galvin Clancey**, and **Gaby Wong**, a team of students determined to create a safe mechanism for groundwater recovery and hopefully win a prize at the same time.

The team was competing in the first annual Designs for a Sustainable World Challenge, hosted by Engineers Without Borders and the University of British Columbia's Sustainability Office.

Student teams were asked to create an object to address a social economic challenge, building it in a short time frame and from what could only be described as garbage. Basically, this was akin to an ultra-sustainable episode of *Junkyard Wars*, with a heavy dose of social responsibility.

The design process for their solution — a human-powered treadle pump — necessitated a serious look at the development challenges in Bolivia, as well as a

survey of the available trash you find in a university setting (lumber, metal rods, plastic piping, etc).

A total of 12 teams were armed with a few power tools and given time to plan solutions. Their tasks varied from increasing peanut-processing efficiencies in Bangladesh to devising ways to lower carbon dioxide emissions in China to capturing fresh water from the misty climes of coastal Ireland.

After frantic planning, a fairly detailed schematic of the treadle pump was produced. On the day of the event, six hours of frantic construction culminated in the final creation. In the end, not only did the treadle pump win first prize, but it also generated the loudest cheer when Wong stepped up on the pump and demonstrated that it did, indeed, work.

"The success of the event and the motivation of the students involved are both living proofs of the desire of today's youth to have a positive impact on the world of tomorrow," says Yifeng Song, one of the event coordinators. And the possibility of a little more fresh water isn't bad, either.

—Dave Ng

Photograph by Yifeng Song

Photograph by Isabel Blas

Eco-Gym

The modern gymnasium is very much a 19th-century creation, no matter how much the fitness freak is kitted out with bad hair, retro headbands, and spandex, or contemporary embedded LCD interfaces and computer-generated body plans. Gyms harken back to a world of classical mechanical physics, plugged into equations of work and energy.

To the strains of Olivia Newton-John's aerobics anthem, the puritan work ethic is transformed into a sweatshop for the body beautiful. The slick machines, treadmills, and cross trainers merely serve to disguise antique apparatuses more at home in a world of steam engines, and to stifle enquiry into thermodynamics and economy.

Then there's artisan **Manuel de Arriba Ares**. Under the sign of his "eco-gym," Gimnasio Ecológico Lumen, Arriba has turned the demon of entropy on its head. Making use of the very waste and by-products of the modern entropic economy, Arriba has created a truly practical monument in the form of a supremely low-tech gymnasium. Its fitness machines, created with a good deal of physical effort over three years from raw and junked materials such as wood, rope, and rubber, directly mirror both the design and functionality of those found within its wasteful counterpart.

Located in the small town of Valdespino de Somoza in the north of Spain, Arriba offers free access for all to this functional work of Art Brut, a wonderful Heath Robinsonesque assemblage constructed from remnants of strollers, boats, bicycles, and automobiles salvaged from neighboring dumps.

Helpful signs, painted on the tarnished white remnants of refrigerators, instruct the would-be eco-gymnast on exercises and operation of the intricate machinery, reflecting Arriba's knowledge and experience over many years as a physical education teacher.

Lumen is a "gymnasium that was born of the nature, (and which) will return to her," Arriba philosophizes. The cycle of waste, embodied by so many aspects of the smogged-out city gym, is closed.

—*Martin Howse*

Real-Life Concept Cars

Like many people of his generation, **Baron Margo** was dazzled by the futuristic concept cars Detroit trotted out year after year. And, like many people, he was disappointed that those streamlined vehicles remained unobtainable concepts to the average motorist. But unlike many people, Margo did something about it. He, as he describes it, "stepped up."

He started to build his own cars. Cars that appear to come from a parallel world, one where you debate whether to vacation on the beaches of Venus or go skiing at Olympus Mons on Mars.

I first saw one of Margo's rocket cars parked at a local diner, a gleaming silver torpedo wedged between unremarkable Corollas and SUVs. Closer inspection showed the work of an incredible craftsman: the sleek aluminum surface was covered with metallic detail, bristling with rivets, lights, and a massive faux jet exhaust with a rotating outer rim.

The three-wheeler uses recognizable parts — a modified front suspension from a VW Beetle, a motorcycle engine — in clever reworkings of proven designs, a practical approach that makes Margo's vehicles not just beautiful to look at, but also legally roadworthy.

But these quite noticeable cars are just the surface. Margo's home is a treasure trove of robots, rockets, and intricate machines, made primarily from found scrap, aerospace salvage, and construction remnants from the Glendale Galleria. Standing in one place, you can see a brass-and-steel train, an old Crosley auto, a gigantic robotic dragonfly, a family of upright robots and their android dog, and so much more. It's dizzying, inspirational, and humbling.

Margo is a reserved man, and while he's sold some works to the rich and famous and to the movies (rayguns for the *Men In Black* series), Margo does what he does simply because he loves it.

Margo is a wildly creative man, a dreamer who manages to actually make things real, thanks to a strong sense of the pragmatic, as seen in the two pieces of advice he gave me: "Take the easiest path" and "Don't burn yourself." Sage advice for every maker.

—*Jason Torchinsky*

≫ **Baron Margo's Cars and More:** baronmargo.com

Photograph by Sally Myers

On the Wings of a Straw

"I remember walking down the beach, looking at seagull feathers and wishing I had a thousand so I could make a kite," explains **Carl Rankin**, a long-time model maker and the man behind several amazing DIY model airplanes.

Rankin's musings about quills led to explorations with drinking straws, and soon, conjoined with thread, tape, and plastic cling wrap, the Jules Verne model was underway. A marvel to watch, this three-tiered, 56"-wide, radio-controlled airplane is capable of flying at walking speed, a feat more challenging than rapid flight.

Ingenious at finding new uses for common household items, Rankin is always on the lookout for new materials. Skewers, cocktail straws, and paper clips have made their way into his models, and one of his favorite inventions is the use of colorful plastic wrap to cover planes like the Jules Verne. It's strong, lightweight, and malleable, with the added benefit of lending surfaces a shimmering glow.

Rankin grew up in a "flying family" and enjoyed making model airplanes even as a child. But back then, he fretted that his models didn't measure up to his brother's. This worry became fodder for an enduring curiosity about accessible techniques and materials for building, and Rankin has since designed hundreds of models.

In 2004, he handwrote a book about one of them, the Foam and Tape Cub. The instructions are packed with valuable and meticulous details about making models, but they have the charming familiarity of learning from a friend. Made from recycled takeout containers and tape, the plain white surface of the Cub practically begs to be customized. Readers from around the world send Rankin photographs of their own versions of the Cub, from brightly painted or fitted with special wheels to inventive double-Cub biplanes.

Perhaps best of all, unlike expensive store-bought counterparts, homemade model airplanes are durable. As Rankin says, "If they crash, you just tape them back together." —*Annie Buckley*

>> **Carl Rankin's Model Planes:** flyingpuppets.org

Photograph by Bob Halvorsen

Hard Wood

In **Lee Stoetzel**'s world, Harley-Davidsons, Volkswagen buses, Macintosh computers, and McDonald's Big Macs all grow on trees. Or at least the materials to make them do.

The Pennsylvania artist recreates iconic products entirely out of wood, with a little steel and Bondo for support. "I stick with very recognizable parts of American culture," he says.

Stoetzel's woody representations tie these objects of conspicuous consumption back to the power of nature and the fragility of the Earth. Just don't call him a hypocrite. All of the wood comes from trees that were already killed by fungus and dredged from rivers. Indeed, it's the wear of the wood that attracts Stoetzel.

In 2004, he built an exact replica of a 1942 Jeep based on an Italeri model kit. The gouges and scars in the pecky cypress wood reminded him of bullet holes, he says, perfect for a classic military vehicle that has since become the quintessential four-wheel-drive.

There's also the 1960s counterculture car-of-choice, the VW bus, whose wood doppelgänger is currently parked in his dining room. That one was tough because he based it on his daily driver. "Every time I saw the real bus in my driveway, I'd notice something wrong about the sculpture," he says.

In the case of *Chopper* (seen above), the custom "Captain America" Harley-Davidson from the film *Easy Rider* no longer existed. So Stoetzel started with a small Franklin Mint replica, scaled up with careful caliper work.

Stoetzel's woodworking chops come on a need-to-know basis. Much of his knowledge was picked up chatting with the "older retired guys" who still put in a few days a week at the woodcraft store where he buys his supplies. The rest comes from the web and DIY books. While constructing the VW, he built his own steam bender from PVC pipe to shape the wood into the bread-loaf shape of the bus.

"After spending two years on the bus, I'm not being as ambitious with scale," he says. "This week, I'm making a pizza."   —David Pescovitz

>> **Stoetzel's Woody Wonders:** leestoetzel.com

Photograph by Rob Carter of Mixed Greens Gallery

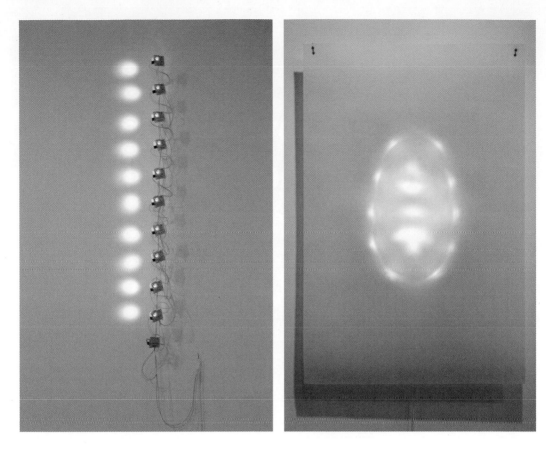

The Nature of Microcontrollers

A Washington native, **Claude Zervas** brings his background as a software engineer to a fine art practice using technology and related apparatuses rather than paint and canvas as a medium.

Zervas' predominant subject matter for the past few years has been the Northwest's verdant and extreme landscape: dense evergreen forests, glacier-melt rivers, and strange roadside attractions.

In his 2005 sculpture *Skagit*, a section of the 150-mile-long river is rendered in glowing green cold cathode fluorescent (CCFL) lamps that climb down from the wall, clamber atop a series of thin steel rods, and eventually split into two forks. Wires and inverters splayed on the floor resemble additional tributaries.

Considering the subject matter, the choice of materials might seem an unusual substitute for the real thing. Zervas' work begs an increasingly important and complex question: what is nature anymore?

In his *Forest* series of computer animations, "forest" is misleading as parts of the landscape have fallen prey to logging. Clear-cuts and swaths of spindly new-growth trees populate the frame until a single-channel computer algorithm set on a continuous cycle slowly morphs and blots the view from existence. Then the cycle begins anew.

"I'm more interested in the memory associations that arise out of perceptions of landscape," Zervas says of his work.

The artist has recently gone from the macro of the forest to the micro of marine life. A new series of wall-mounted sculptures uses what Zervas calls "motons" (small circuit boards studded with alternating blinking lights run by a microcontroller) to investigate the phenomenology of simple life forms.

Their movement is so quick that it's hard to tell anything is happening at all. What the brain registers instead is the space between — similar to how Zervas situates the viewer between dying landscapes and new technology.

—*Katie Kurtz*

≫ **Claude Zervas:** claudezervas.com

"Since I discovered I'm an RX680, women definitely find me more attractive.

Well, at least in photos."

▸ **Everyone's got an Epsonality.**
Discover yours at Epsonality.com

▸ **Todd**
Account Supervisor – Seattle, WA

Epsonality Type:
RX680

The Epson Stylus® Photo RX680. Performance photo printing with all the bells and whistles. And an LCD preview screen. And auto two-sided printing. And two trays for double paper capacity. It's the all-in-one for Ultra Hi-Definition Epsonalities.

Make Free

THE GREAT, PERVERSE JOY OF STEAMPUNK

By Cory Doctorow

TOM JENNINGS IS A SKINNY, PIERCED, and tattooed DIY engineer in his early fifties, a self-described "queer punk." In 1984 he launched FidoNet, a way for dial-up bulletin board systems (BBSes) to pass messages back and forth. FidoNet let each BBS spool up messages for other BBSes, then phone in whenever rates were cheapest to pass them on.

FidoNet nodes passed messages on to other nodes to be delivered further down the line, in a robust and elegant fashion, so that messages could be delivered to almost anywhere in the world for free. (The alternative? Telegrams, expensive long-distance voice calls, or leased lines).

I first met Jennings long after the founding of FidoNet, at a conference in a remote location near Silicon Valley. He was putting on a performance, a uniquely technological, Jennings-esque show. A MAKE contributor, Jennings hoards and restores antique computer equipment, and that weekend, he showed up with a bunch of bar-fridge-sized tape readers and teletype printers, sheet steel devices that smelled of machine oil.

He'd prepared a biography of computer science pioneer Alan Turing, punched into a long roll of black paper tape. His homebrewed tape reader had no uptake reel — the tape went through the reader and unspooled onto the floor, all weekend long, until it made a waist-high, snarled mountain of paper. The reader was connected to one of those fridge-sized printers, and all weekend long, it printed out the biography. When it was over, Jennings threw away the tape.

That solid bar-fridge of a printer, like the walnut, waist-high console stereos of my grandparents' rec room, hearkened back to a gentler era, an era when technological change unspooled at a more genteel pace. Back then, it made sense to build a machine to last for ten years, 20 years, 50 years. Machines of all kinds — plows, log splitters, printers, stereos — were things of enduring utility.

Today, most machines aren't built to last. iPods are finished with material that scratches if you look at it sideways — put it in your pocket for five minutes and it looks like it's been in a rock tumbler for a month. It's not as though we don't know how to make alloys that resist scratching; I have a pocket full of change that's been grinding against keys for years and still looks sharp.

But what would be the point of an iPod built to withstand the ravages of time? Remember those gen-one MP3 players, the size of a half-brick, able to hold a whopping two or three gigs? Keep one of those for ten years and it becomes an unwelcome houseguest, a relative who's overstayed his welcome. You want to be able to toss that overweight, underpowered brick in the trash ten to 18 months after taking it out of the package without feeling the guilt that comes when you set a well-made, precision-engineered console hi-fi out at the curb, still smelling faintly of the Murphy's Oil Soap your family's been rubbing into it since 1965.

This is, I think, the great, perverse joy of steampunk DIY projects — taking disposable junk-tech intended to last for a year and putting it into an artisanal enclosure that stands in testimony to the artifice of the cabinetmaker, the metalworker, and the leathersmith who worked together to turn this ephemeral technological moment into a lasting statement that could find its way into a museum someday.

What does it mean to live in a high-tech era and be fascinated by the aesthetic of steam, the neo-Victorian maker look? It's neither futuristic, nor particularly nostalgic; rather it tells us that today, in the present moment, we are practicing the mental gymnastics necessary to find something beautiful and exciting today and disposable tomorrow.

Cory Doctorow (craphound.com) is a science fiction novelist, blogger, and technology activist. He is co-editor of the popular weblog Boing Boing (boingboing.net), and a contributor to *Wired*, *Popular Science*, and *The New York Times*.

Art and Culture

Philip Ross' art crosses the boundaries of technology and biology.

By John Alderman
Photography by Jonathan Sprague

FROM HIS STUDIO HIGH ON A HILL ABOVE SAN Francisco, artist Philip Ross has a firsthand view of the transformation, expansion, and human activity below — a window onto architecture and ecological shifts, whether they be human versus earthquake over 100 tumultuous years, or puffs of fog hydrating backyard plants on a lazy afternoon.

Within the studio are the artifacts of Ross' own shifts, the mulchy, calcified, glassy, wooden, digital, and pulpy remains of more than a decade of art-work that straddles the border of technology and biology. Sometimes wrapped neatly, sometimes just lying on the floor, these objects map Ross' movements along a path where aesthetic curiosity has led directly to a hacking of scientific methods.

Ross' work explores our inescapable roots in a constantly shifting biological world. It's a field with easy appeal: few things are as startling as life in transformation. And it's no accident that his art overlaps with science. The notion of the Greek word *techne*, as Ross explains it, allows for discovering the properties of a thing not just by what it appears to be, but by what it might be, what it can be: "Both art and technology look at the world like that — not necessarily as a given but as potential, possibility, directions, alternatives."

Ross himself is a hybrid — born in New York, transplanted to San Francisco, and not quite com-fortable with any particular label. Artist, teacher, and curator, Ross is a scavenger as well as sculptor of living objects. His personal quest to engage with, and teach others with, the fundamentals of life fits perfectly with our time, when fields like biology and engineering are merging and raising questions that call for a keener sense of responsibility for where we're taking this planet.

"My agenda is partially an education," Ross says, at breakfast, pausing over pancakes at a rowdy Mission District diner. "We have to move beyond a naive ecological view of where we are situated, in that we are definitely animals and our technology is a natural part of our condition, and if we don't figure out how to naturalize our technologies or get them in line with nature, that's where the real problem is."

The pieces Ross makes often place living forms in a situation where human involvement is formalized, like *Junior Return*, a work from 2005 that encases a plant in a blown-glass hydroponic pod, with digitized life support dialed to the absolute minimum.

The result is a stunted, live object of rarefied beauty suspended in an eerily lit, tiny stage. (Ross recently reprised this work with a batch of 18 pods, all networked together and powered by central batteries.)

The body of work is as much about process as outcome — or at least that's where the challenge lies. For *Was Below, Now Above* (2002), Ross created a steel frame that he covered with very tiny oysters and submerged for two-and-a-half years, allowing a colony to form and mature.

As the day approached to raise up the steel ribbing, Ross knew that if he wasn't careful he'd soon have a ton of rotting mollusk meat on his hands — a hazard to his piece and possibly public health as well. Finding out how to safely and quickly rid the frame of the meat was a challenge that led him to universities and museums, and finally to some not-so-salty water, where the change in salinity killed the oysters, and bacteria ate the soft tissues but not the shells, leaving stunning skeletal remains.

Then there are the reactions from viewers. While building an installation for the Exploratorium in San Francisco, Ross learned that one out of every 1,000 children is a "whacker," or someone who will just

Ross in his studio, overseeing his work that links hydroponic pods of willfully stunted plants into a challenging demonstration of control and life's ability to adapt.

instinctively bash an installation. With *Junior Return*, comparisons with bonsai are apt, and frequently made, but what's remarkable is how often parents project onto the work concerns about their own caregiving, or how office workers recognize a metaphor for their own highly regimented lives.

Ross wants to address those issues, and also something a little more fundamental in how we think of ourselves. "It's thinking about how we actually form the Earth, and where we are in the waste cycle, or in the consuming cycle. It's part of our 'animalness' that we don't just own materials or solidify materials — they pass through us in specific

"You can just sit and watch the grass grow and come up with phenomenal postulations about the nature of life. A notebook is pretty much all that Darwin had, and he did pretty well."

channels or streams. But we don't think like that, or think of ourselves in biological terms. We think of ourselves in terms of artifacts."

Not always, of course, and one is reminded of Buckminster Fuller's famous proclamation, "I seem to be a verb." Ross' work follows up on Fuller in attempts to conjugate that verb in all its tenses. A willingness to ignore categories in favor of following what seems most interesting is something Ross seems to have picked up on the West Coast.

AFTER AN ADOLESCENCE SPENT AS A self-proclaimed "straight-D student" in New York City, Ross moved to San Francisco to study at the San Francisco Art Institute. The Bay Area's thriving technological experimentation was inspirational, with hotbeds of activity stretching from industrial artists like Survival Research Laboratories to corporate innovators like Xerox PARC.

"Compared to the East Coast it's a lot easier to integrate these different worlds; people cross

over. The Bay Area not only has this great technical culture, it also has a technological counterculture." Technophile splinter groups with "thousands of opinions" expanded Ross' sense of possibility.

But a fascination with technology was a long way from the kinds of biological experiments that Ross currently produces. Perhaps predictably, it was his day jobs in other professions that led him to the natural world.

"Mushroom hunting unlocked it all for me," says Ross, who stumbled upon his obsession in the days after art school when he supported himself as a cook at a summer camp. He loved to eat, and when taken along on a hike with a fellow cook, he found that searching for his own food in the wilds offered a much different sense of nature than the alienation from it that he felt growing up.

"It was a way that I could also pay attention very critically to my environment and the subtle things going on, because it's so high-risk: you do the wrong thing and you're dead. You spend time learning about this thing or find experts who you trust. Who do you trust with your life?" His interest in mushrooms led to learning about the trees they grow alongside, and from there the expansion was infinite.

Ross changed jobs, this time to become a hospice caregiver. Grueling work, its constraints wouldn't let him go out in the woods mushroom hunting, so he became determined to teach himself how to culture and grow them at home, reading up on scientific protocols. Though they used a different language, the protocols for mushroom growing resonated with his cooking experience. "The tools were all the same, and even the procedures: steaming, and baking, and thinking about cleanliness." It all came down to having a recipe.

Once he realized that working in a lab was like cooking, he jumped right in and started building his own laboratory (see MAKE, Volume 07, page 102, "Home Mycology Lab"). "It wasn't too complicated. Doing biological research is really easy. You're dealing with biology, and it's everywhere around you."

Ross' most ambitious work yet debuted in late October with BioTechnique, an exhibit that he

A sketch for a fungal re-creation of Harold Edgerton's famous *Milk Drop Coronet* photo stands over the collection of tools and artwork that blend together in Ross' studio.

is curating and also exhibiting in. Held at San Francisco's Yerba Buena Center for the Arts, the show assembles a group of artists whose work incorporates biology, alongside works from industrial technologists, ecological researchers, and biological engineers.

Fitting everyone together in a museum is "like a temporary hospital-space-slash-research-lab, in a setting that doesn't really accommodate that."

Art museums are just not comparable to, say, a university's ability to pay the overhead for all this science. But Ross' ability to network, ask for help, and call in favors from his friends and allies, "taking little bits from all over, volunteering from all around, and stitching it together," will allow works such as the one planned by Australia's Tissue Culture and Art Project, that will involve growing out a specific mouse cell line. The "McCoy" cell line will need to be ordered, cultured at the nearby Exploratorium, and then transferred to the museum and kept in a bioreactor.

There's also a concurrent educational event tied to the show, called Technebiotics, a day of demonstrations of biological techniques by artists, scientists, and other show participants.

"I love showing people how easy it is to build all this stuff," Ross says. Teaching art at Stanford and UC Santa Cruz, he proves that the line between artist and scientist is very slim. The barrier of having the right tools, he demonstrates, is low indeed, with exercises like "Medieval DNA Extraction" in which he shows students how to extract DNA from mammalian or vegetable cells, using candles as a heat source, some water, and not much else. "In the end you have this vial in front of a candle, and it's almost an alchemical thing: suddenly you have DNA."

The thinking is the important thing, and the willingness to try. "Almost all biologists that I know, their greatest skill is observation and comparison and very good data acquisition. You can just sit and watch the grass grow and come up with phenomenal postulations about the nature of life. A notebook is pretty much all that Darwin had, and he did pretty well. Anybody can engage."

➕ Philip Ross' artwork: philross.org

John Alderman (john@supereverywhere.com) is a writer and user experience consultant who lives in San Francisco and travels often, for work and whimsy. His recent book, *Core Memory*, was featured in MAKE, Volume 10.

Floating City

BELGRADE'S GYPSY HOUSEBOATS ARE THE DWELLINGS OF THE FUTURE.

By Bruce Sterling

NATIVE TO BELGRADE'S TWO RIVERS, THE Sava and the Danube, the *splav* is made from repurposed industrial junk. This raffish watercraft probably owes a large design debt to Belgrade gypsies, who are commonly scrap metal dealers. As Eastern Europe's rock-bottom underclass, many gypsies literally live inside Belgrade junkyards, in fantastic sheet metal huts wired together out of anything handy.

So how do you make your own free-living splav houseboat? It's intriguingly simple! First, get your hands on some empty German chemical-industry barrels — hopefully these drums held something nontoxic, such as paraffin or corn syrup. They're cheap or free, since huge barges full of German oil drums steadily ply the Danube. Then scare up some rusty, Communist-era angle iron from any of the city's many junkyards.

Take the German drums and the Yugoslav iron to a local marina at riverside. Weld the angle iron into a tight box that traps 12 or 15 of the watertight drums. Throw down some cheap wooden flooring over this buoyant iron foundation, then nail up a frame and a roof, doors and windows.

Launch your splav and have it towed. Find some spot on the shore that seems unclaimed, pound in a literal iron stake, tie a hawser to that, and you have staked your claim!

Further refinements are entirely up to you. Hanging dead rubber tires off the rim of your splav is a cordial touch, since friends with boats may visit, and these fenders will spare their hulls.

Of course you will have no mailing address, but you may be able to steal some electrical power, as that's not well policed. Flowered window boxes are a gracious touch. If you're a migrant fresh from the village, or a displaced Balkan refugee, you might also plant a little subsistence garden on the shore. Nobody will stop you from fishing the kindly Danube, and even if you're one fish-and-pepper soup away from utter destitution, everyone will think you are amusing yourself like a city gentleman.

Building a houseboat on top of floating oil drums is extremely simple and cheap. That's why there are hundreds of *splavovi* — more every week. But how do people get away with doing it?

Seattle's once-colorful Lake Union houseboat village, which dates back to the 1890s, has long since been gentrified, Microsoftie-style. Amsterdam's houseboats are legally fossilized in amber. Belgrade's splavovi are much more modern animals. They evolved in an outlaw mini-state with a post-Communist economy in a turbulent transition. This means, as a hands-on vernacular architecture, they have a uniquely favorable economic, political, and legal environment.

This richness of creative opportunity has caused the splavovi to break up into several different species.

Basic Splav This is poverty slum-housing, over water. Somebody's trying to make a full-time go of life on a splav, for lack of other choices. They may be refugees, smugglers, artists, drunks, poachers, visionaries, migrants, or retired on a nonexistent pension. Their lives are pretty hard, and they look it.

Recreational Splav This one was thrown together as a cute, toy, floating summerhouse by someone who enjoys the river. Rarely visited by their owners, who generally have real jobs and other homes, these splavovi often mildew or catch fire.

Speakeasy Splav This is what happens when the inhabitants of the Basic Splav catch on to the unique legal advantages of their situation. No taxes, no licenses, no fire-safety codes ... no identity. Why

not sell booze on board, or any other contraband that someone might like to consume? Yo ho ho!

Nightclub Splav Given that you're already running an illegal bar, why not scale it up radically? Bring In a gypsy band, some techno DJs, or the fiercely popular "turbo-folk" ethnic divas! And — given that mere oil drums can only support so much dance floor — why not grab an entire dead barge or outdated river ferry, and just nail that hulk into place on the riverbank? You'll have to settle that feat with the local cops somehow — but hey, cops are husky young guys, cops love boozy, miniskirted Belgrade party girls!

Deluxe Restaurant Splav Since everybody's partying on the river anyway — life sure feels easier there, much less constrained somehow — why not feed them, too? The view of the Blue Danube is hard to beat, the air is cooler, ships have galleys — and if they don't, you can always carve out a kitchen with a blowtorch.

Nouveau Riche Splav The urge to show up the neighbors is as old as mankind, so why not a big, gaudy, two-story splav with aluminum siding and maybe the tasteful stylings of a fake Chinese pagoda? Given that there are so many splavovi,

Splavovi are inexpensive houseboats that float along the Sava and Danube. This drydocked one still has the oil drums.

a host of handymen are in business making them, and they're looking less spontaneous and more like a houseboat industry.

Over the past few years, Europe has been learning a lot about climate change. The summers are much hotter (that's good for splavovi, because the river is cooler than anyplace else in town). The winters are milder (also good for splavovi, because they are flimsy, lightweight, and poorly insulated). The droughts are longer (not too bad, as splavovi can squat in the river mud if need be) and the rains are sharp, sudden, deep, and sometimes catastrophic. The flood of 2006 took a toll on splavovi.

So the trend is clear: both rivers' shores will end up with huge, wet margins of uninsurable, semi-habitable, climate-change slums — the new urban river marshes. Such a severe situation will take a lot of painful adaptation. Unless you're halfway there already.

A river isn't a toy for city dwellers. A city is a toy for a river.

Bruce Sterling (bruces@well.com) is a science fiction writer and part-time design professor.

Building the Barrage Garage

The first in a three-part series about building a workshop from the ground up. By William Gurstelle

AS A CITY DWELLER, I'VE OFTEN LOOKED with envy at the spacious outbuildings of my rural friends and relatives. Horse barns, potting sheds, root cellars, equipment garages — plentiful, enclosed, and private space is the one thing that makes me envy those who live beyond the end of the bus line. I think often about what I could make if I had a room of my own: a purpose-built, well-equipped space in which to create.

Apparently I'm not alone in these thoughts. Homebuilders commonly offer two-, three-, and even four-car garages for new homes. But all that space isn't needed simply to shelter the family Chevy. It's needed to keep pace with the explosion in DIY projects and their concomitant material and tool requirements.

Randy Nelson, president of Swisstrax, a manufacturer of workshop and garage floor products, says that garages are quickly evolving into more than simply places where people keep their cars. Installation of the company's special-purpose floor tile in garages and workshops is booming.

Photography by William Gurstelle

"[Spaces for making things] have just about doubled in the last ten years," says Nelson. "People aren't just stuffing junk in their garages any more. It's become the male domain, the place where they can do their work and have their tools."

There are scores of books providing advice on setting up a wood shop or metal shop, and many others that describe setting up specialty areas: a paint shop, a photography studio, or a chemistry lab. But what I wanted was not a single-purpose workspace. I was seeking the ultimate, multipurpose Maker's Workshop: a versatile, flexible space capable of handling nearly any project I could think of — from building a cedar-strip canoe to compounding fuel and oxidizer for a rocket engine, from soldering a Minty Boost to developing a model ornithopter.

This article is the first in a series detailing the creation of a modestly sized yet state-of-the-art Maker's Workshop, which I named the Barrage Garage. This installment covers the design and construction of my Barrage Garage, and the considerations behind its climate-control systems, floor coverings, and other infrastructure. Future articles will describe the equipment inside it, such as workbenches, machine tools, hand tools, and welders.

Workshop Design Criteria

The first step was to determine which features were the most important and practical.

Egress A 9-foot-wide, automatic, well-insulated door outfitted with required safety equipment was essential. The huge door makes bringing materials in and out of the workspace a snap.

Fenestration Natural light and a view to the outside were high on my list of priorities. Therefore, the design called for four east-facing sliding windows having a total glass area of 24 square feet.

Organization I devised a plan for a combination of stackable modular cabinets, which, along with a slotted wall storage system, maximize the efficiency and versatility of my space.

Surfaces I wanted more functionality and style than a concrete floor could afford. I selected a special-purpose tile floor for workshops and garages that makes walking and standing more comfortable.

Power I needed 240 volts to run the heater and welder, and 120-volt receptacles placed at frequent intervals along all walls on two separate 20-amp, GFI-protected circuits. This ensures a plentiful, safe supply of electrical power to all tools.

Building the Barrage Garage

My first task was to site the structure. Where should the workshop go?

Initially I considered placing the shop in my basement. Possible, but this would involve far too many compromises. The basement is a low-ceilinged space with marginal access via a narrow stairway. The thought of carrying tools and materials up and down, turning corners, and so forth quickly dissuaded me.

Instead I turned to the nearly forgotten space along the alley in back of my home. Separated from the rest of my yard by a chain-link fence, it was covered with 25-year-old lilac bushes. I loved those fragrant, beautiful spring blossoms, but the space those lilacs grew upon was workshop-perfect: it had room, privacy, and access. So, goodbye lilacs.

City ordinances allowed me a maximum of 240 square feet for the shop. With the city building permit obtained, it was time to push some dirt.

PUSHING DIRT

It all starts with a level floor. Every workshop, atelier, pole barn, or garage must have a level floor if great things are to be made in it. It has always been this way. Four thousand years ago, in the reign of the great Egyptian pyramid builders, construction techniques were rudimentary. Imhotep, legendary architect of the pharaohs, had only knotted measuring ropes stretched taut between stakes, plumb bobs, and sighting sticks.

But Imhotep gave the pharaohs the tools to build monuments capable of withstanding 50 centuries of desert sandstorms. He did that by starting with a perfectly level floor. It's believed that the Egyptians leveled the area under a pyramid by cutting a shallow grid of trenches into the bedrock, then filling them with water. Knowing that the height of water within connected trenches would be at exactly the same level, the workers hacked out the intervening islands of stone and sand with hoes and stone drills.

The Barrage Garage has a flat floor as well, but my excavators used a 75-horsepower backhoe and modern surveying tools including transits and lasers. My end result is pretty much the same as Imhotep's: a perfectly level slab placed in exactly the right spot.

CONCRETE IDEAS

After excavation, the concrete work began. Concrete is composed of Portland cement, gravel, sand, and water. When freshly poured, concrete is wet and plastic. But within hours it begins to solidify, ultimately becoming as hard as rock.

Most people call that process "drying," but the concrete crew foreman on my job told me that's not really the best choice of words. Concrete does not simply solidify because excess water has evaporated from the slurry. Instead, the water reacts with the cement in a chemical process known as hydration. The cement absorbs the water, causing it to harden and bond the sand and pebbles together, creating the stone-hard material we know as concrete.

FRAMING THE CONCEPT

Prior to the mid-19th century, building was an art that took many years of apprenticeship to learn. There were few if any building codes. Quality of work was based largely on the personal integrity and craftsmanship of each builder.

For 2,000 years, the most common technique for building with wood was the method called *timber framing*. Buildings of that era still exist; typically they are barns and homes with huge wooden beams supporting large open spaces.

In the mid-19th century, building techniques changed. Cheap, factory-produced nails and standardized, "dimensional" lumber from sawmills allowed for a faster, more versatile method of construction called *balloon framing*.

Invented by Augustine Taylor of Chicago, balloon framing revolutionized building construction. It utilized long, vertical framing members called studs that ran from sill to eave, with intermediate floor structures nailed to them. What used to take a crew of experienced timber framers months to join and raise, could be constructed in a fraction of the time by a competent carpenter and a few helpers.

Over time, balloon framing evolved into the current technique known as *platform framing*. The Barrage Garage, like most modern buildings, is built by nailing together standard dimensional lumber — 2×4 trusses holding the roof and 2×6 studs forming the walls — at code-defined intervals. Then, plywood sheathing is attached to the lumber frame, and the basic structure is complete.

A SOLID FLOOR

The first order of business after the workshop shell was complete was to install the floor. There are three general options: coatings, mats, and tile. Each has its own advantages and disadvantages.

Most common and least expensive are coatings. There are several types of coating available for concrete floors, including epoxy, polyurethane, and latex.

Epoxy paint is probably the most widely applied form of floor coating. Epoxy forms a hard, durable surface and bonds solidly to a correctly prepared surface. Because floor coating provides no cushioning, it can be hard on feet and legs. Also, it doesn't last forever: expect to recoat the floor every five years or so.

Polyurethane coatings are also very durable, and they resist chemical spills better than epoxy. But urethanes do not bond directly to concrete, so an epoxy primer coat is required.

Latex garage paint is widely available and inexpensive. It goes on easily and doesn't require the prep work associated with epoxies and urethanes. However, it is less durable.

PVC floor protection mats are another option. They protect the porous concrete floor from staining or corrosive chemicals such as oil, paint, or acid. Mats are typically simple to install, requiring only scissors. Importantly, they add a cushioning layer above the hard concrete.

Special-purpose vinyl tile is the premier flooring option for workshops and garages, and that's what I installed in the Barrage Garage. These floor tiles, from Swisstrax (swisstrax.com), snapped together firmly and were easily installed without special tools.

Tile handles heavy loads and high traffic. It resists damage caused by chemicals, and it's far more comfortable to stand on than concrete. But best of all is tile's ability to transform a humdrum workshop into a great-looking space.

Tiles come in a wide variety of colors, which allowed me to create my own floor design and inspired me to echo the floor colors on the walls and window trim, and in a cool wall-mounted atomic ball clock and coat rack inspired by George Nelson (*see CRAFT magazine, Volume 01, page 135*). Now that's a workspace designed to inspire a maker!

William Gurstelle is a MAKE contributing editor.

Gas or Electric Heat?

It gets cold (and hot) in Minnesota, where I live. Therefore, the workshop must be well insulated, heated, and ventilated.

The walls are framed from 2×6 studs, which provide enough depth for R-19 insulation, and the windows are double-glazed.

I briefly considered a hydronic radiant floor heating system. A hydronic system uses in-floor, hot-water-filled tubing to heat the room from the ground up. I knew that the concrete slab floor would be cold, and that a hydronic system would make the room warm and comfortable. But the hydronic system has a longer heat up time and much higher initial costs. This swung the decision toward a fan forced-air heater mounted in the rafters.

But is gas or electric a better choice? A good case could be made for either. Deciding which made more sense was a study in what my old college professors would term engineering economics.

Step 1. Determine operating expenses.

To make a solid financial decision, I needed several pieces of information: the relative energy costs for gas and electricity, the purchase costs for each type of heater, and the efficiency of each.

I studied my utility bills to determine comparable energy costs. January natural gas costs me about $1.10 per therm, a therm being the energy equivalent of roughly 29 kilowatt-hours. So $1.10 divided by 29 equals a gas heating cost of 3.4 cents per kilowatt-hour. The local energy utility charges 7.2 cents per kilowatt-hour.

Electric heaters are 100% efficient, that is, all the energy input goes toward making heat. Gas heaters are about 80% efficient. To account for the gas inefficiencies, I divided the gas cost of 3.4 cents/kWh by 80%, yielding a net cost for gas heat of 4.2 cents/kWh. Bottom line: where I live, gas is 3 cents per kilowatt-hour less expensive than electricity.

Step 2. Determine initial costs.

So, the smart money goes with gas, right? Well, not necessarily. While comparable-sized gas and electric heaters cost about the same, gas line installation can be expensive. The contractor quoted a price a bit north of $1,000 to trench, plumb, and install a gas line to the workshop.

Step 3. Compute the payback period.

$1,000 divided by $0.03/kWh = 33,333kWh

The heater I've chosen operates at a maximum of 5 kilowatts. So, each full hour of use costs 15 cents more if I use electricity instead of gas. Dividing $1,000 by $0.15 per hour computes to 6,666 hours of heater use. Given the relatively light use of the shop, the payback period on the installation of a 5kW gas heater would be decades, so I invited Reddy Kilowatt into my shop.

NEXT: *In Volume 13, Gurstelle explains shop organization.*

CandyFab

How we built a 3D freeform sugar printer in our kitchen.
By Windell Oskay and Lenore Edman

LAST YEAR, WE HAD A CRAZY IDEA. It started out something like this: "Wouldn't it be cool if we had a 3D printer?"

Machines for printing three-dimensional forms do, of course, exist, and utilize a variety of complementary and competing technologies known by such intimidating names as stereolithography, selective laser sintering, and fused deposition modeling. If this were ten years in the future, we would just go down to one of our local big-box stores and pick one up.

Unfortunately, solid freeform fabricators, as these machines are collectively known, are not household tools; they almost universally come with five- or six-digit price tags. They are intended for rapid prototyping, making precise models of parts that otherwise would be made by expensive and time-consuming conventional machining processes.

But we were really after a device to play with, not some uber-expensive industrial manufacturing system. So we built one ourselves.

Our idea was to use sugar as the printing medium. It's remarkably inexpensive and is rigid despite its low melting point. As it melts, it gives off heavenly fumes — the final touch making it ideal for the home prototyper. To actually build something with it, we came up with a low-cost, low-tech method that we call Selective Hot Air Sintering And Melting (SHASAM — good acronyms help). We move a hot air gun over a bed of sugar to melt it locally. The bed is then lowered, another layer of sugar is added, and the heater is moved over the new layer, melting

LEFT: A toroidal coil sculpture printed out of pure sugar on the CandyFab. THIS PAGE: The 24"×13½"×9" build volume of the sugar printer is a rectangular pit in the middle of the canvas liner. While printing, the two stripped-down pen-plotter mechanisms move the hot air gun over the bed of sugar in the pit.

deeply enough to fuse to the layer below. As the process repeats, true 3D objects are produced.

We started with two junky old pen plotters that we stripped down to their useful parts: quadrature-encoded motors, belt drive systems, and sturdy mounts. To these we added an electric automotive jack (for the vertical motion of our heavy box of sugar) and a plywood frame. The tricky part of this process was designing the plywood parts — and the overall scale of the machine — to fit the found plotter parts. Past that, we added a flexible canvas liner to keep sugar from leaking out, and micro-controller-based drive electronics. Our low-velocity hot air gun "print head" was made from parts including the heating element from a soldering iron and a small aquarium pump that provides airflow. With a few late nights and a lot of luck, this crazy idea actually worked.

Since we got it working, we've used our new fabber to print geometric sculptures, a flexible chain of links, and a sugar bowl — made of sugar. We've even decorated a cake with fabbed snow-flakes. More recently, we've been experimenting with printing plastic and have made real progress in improving the precision of the machine.

Beyond mere refinements, we're taking this project

We were really after a device to play with, not some uber-expensive industrial system.

open source. To that end, we've launched the Candy-Fab Project at candyfab.org. Its goals are to reduce the costs associated with solid freeform fabrication, and to promote the use of fabrication technologies for culinary, educational, and artistic purposes.

The opportunities for participation are diverse: sewing, programming, gastronomy, and materials science are just a start. We are redesigning the machine to use off-the-shelf components controlled by collaboratively engineered hardware and soft-ware. Project participants are already trying out new designs and researching materials. With all this help, we plan to release plans for a new and improved model next year, so that everyone else who wants a 3D sugar printer can build one too.

Windell Oskay and Lenore Edman have a website called Evil Mad Scientist Laboratories. evilmadscientist.com

For Whom the Bell Tolls

David Gurman's installation rings bells to make people aware of ongoing military tests hundreds of miles away.
By Katie Kurtz

Divine Strake Project, a sculptural installation by Bay Area artist David Gurman, is an intricate, collaborative project that addresses several complex issues: institutional land use, military strength, and unseen forces that have the power to irrecoverably shape our lives. The project consists of bells that are rigged to chime when software detects seismic activity at the Nevada Test Site, a desert facility where the U.S. military has conducted nuclear weapons testing since 1951. Named for an actual large-yield, nonnuclear, high-explosive test slated to happen earlier this year but since halted, Gurman's *Divine Strake* is a hybrid of new and old technologies that addresses the complicated fact of war. Exhibit dates and locations for 2008 are available online at davidgurman.com.

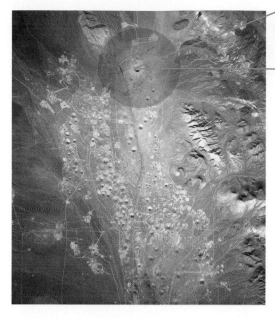

"I thought the bell as a beacon and a call to worship was an ironic way to signal nuclear activity."

What was the initial impetus for *Divine Strake*?
I was spending a lot of time in the Great Basin Desert and learned about the Divine Strake test. [This high-yield underground test has since been cancelled, due in part to a public outcry to stop it.] So far, 1,258 nuclear detonation tests have been performed at the Nevada Test Site to date, creating these massive craters as large as 3,000 feet across and 600 feet deep. A lot of people don't realize that subcritical nuclear weapon testing is still happening. I wanted to create a sculptural installation that would allow people access to something they may know exists but can't readily see.

What is the signal, and where does it come from?
Seismologists have set up arrays at the Nevada Test Site to track potential violations of the Comprehensive Nuclear-Test-Ban Treaty. *Divine Strake* is set up to receive data from the University of Nevada, Reno array. Real-time data is streamed to a network that uses a seismic monitoring software package, Antelope, donated by Boulder Real Time Technologies. A reasonably tested origin and magnitude are furnished to Antelope, which sends it to a Max/MSP [environment] that triggers switches and turns on the actuators to strike the bell.

Based on the coordinated information, you can only speculate whether it's a weapons test or latent geological seismic activity. What I'm interested in is how the information is obscured and that the power is held in the potential threat — the knowledge itself as opposed to the actual event.

Tell me about the bell.
I thought the bell as a beacon and a call to worship was an ironic way to signal nuclear activity. The bell was cast in 1908 at the Andrew Meneely and Benjamin Hanks foundry. I wanted to use a bell from a foundry that had ties to the revolutionary era when Manifest Destiny and westward expansion were set in motion. This particular foundry opened in 1826 and cast cannons in wartime and bells in peacetime, so all these histories are embedded in the very alloy of the metal.

How is the bell rung?
I designed a robotic arm to resemble a gross appendage as a way to highlight the digital age uniting with an analog historical object. The arm is made from water-jet-cut aluminum, linear actuators, and four tolling hammers. A Teleo module receives a signal from the computer to turn on one of the four actuators. Each one has a designated seismic threshold: the smallest registers 0–1, the next 1–2, then 2–3, and the largest 3+. The sound bow near the bottom and thickest part of the bell is the lowest frequency. As it moves up the tollers and hits the vertical axis, the tones are higher, which coordinates to smaller seismic events.

What's next?
The ultimate manifestation is to monitor weapons testing on a global scale and to have the piece sited at various international venues. Bells would be hooked up to register seismic activity at every nuclear weapons testing site in the world, such as North Korea, Russia, France, China, and so on. When the bells chime, it won't be clear which country it's coming from.

Katie Kurtz writes art previews, art reviews, and artist profiles for various publications. She's currently working on a multi-genre project called "Dreamland" about the atomic bomb, dreams, and love.

Escape of the Blubber Bots

Jed Berk's autonomous blimps are on the move. By Mark Allen

BERK'S BLUBBER BOTS KEEP TRYING to escape. During an exhibit in an empty cold-storage space, one wandered out the front door never to be seen again. At a recent workshop in Los Angeles, another made a slow break for the back stairs.

These elusive Blubber Bot entities are autonomous robotic blimps. Roughly 3 feet long, their helium-filled mylar balloon bodies are propelled by ultralight motorized fans. Simple light sensors send them in the direction of the brightest light source in view, and a set of bump sensors tells them to turn around if they run into something.

Blubber Bots also enjoy "networking" while drifting around gently bumping into things — an off-the-shelf detector of cellphone signals causes them to whirl around and produce tones inspired by the songs of whales. Wave your cellphone at one to say hello! In a space with a ceiling high enough for a pack, they can drift about almost unnoticed, like a sneaky group of cartoon clouds.

Blubber Bots are one of Jed Berk's several projects inspired by natural life forms, which he calls *TransitionalSpecies*. These sculptures possess simple behaviors, emulate interspecies communication, and model processes of biological systems.

Photograph by Syuzi Pakhchyan and Sebastian Bettencourt

In groups, they create "biotopes" — environments where people can become part of a temporary ecosystem with these new electronic organisms.

Berk's interest in working with art and technology started while attending graduate school at Art Center College of Design in Pasadena. While working together in a class taught by Bruce Sterling, Berk and fellow student Nikhil Mitter collaborated with architect/software designer Ewan Branda to develop Autonomous Light Air Vessels (ALAVs), the ancestors of the Blubber Bots.

Based on Sun Microsystems' wireless micro-controller development system Sun SPOT, the ALAVs grew out of an interest in matching independent robots with compact wireless communication technology. ALAVs use a node-based wireless network to travel in flocks, communicate with people using voice recognition, and exhibit a sophisticated range of behaviors including feeding, scattering, and courtship. One of the more intriguing features is the "happiness factor" — the flock's altitude, sounds, and light activity depend on interrelated variables that constitute their general sense of well-being.

If the ALAVs are a school of networked dolphins, the Blubber Bots are more like a pack of napping manatees — drifting along toward light sources and turning around if they bump into something.

Developed with artist/biologist/systems designer Bruce Hubbard, the activity of a pack of Blubber Bots is a model of emergent behavior, the process by which complicated or unpredictable patterns can emerge from simple tropisms, such as being attracted to light or turning away when bumped into. Pack behavior emerges on its own, without the need for communication between units.

In response to the many inquiries he received after exhibiting the ALAVs at conferences and art exhibits and on the Discovery Network, Berk developed a blimp kit that anyone with a soldering iron can put together. The current version is available B.Y.O.H. (Bring Your Own Helium) from the Maker Store (store.makezine.com).

THIS PAGE, LEFT: Kids investigate a blubber bot as part of a Build a Robot program at the Art Center for Kids this past summer. RIGHT: Jed Berk and his ALAVs at the Robots at Play 2007 Festival in Denmark.

In addition to the basic Blubber Bot, Berk plans some interesting branches of the family tree. One version uses all surface mount components and will be the size of a standard party balloon — all the better for casually infiltrating living rooms and small apartments.

Another version in development with USC professor Julian Bleecker will provide code modules for Arduino, the microcontroller development environment used to program the brains of the Blubber Bots. With a set of publicly available code modules, programming your own behavior for the Blubber Bot will be much easier, and the Blubber Bot species can continue to evolve new ways of relating to the world. Berk and Bleecker hope this will be a big step in the creation of an open source platform for blimp robotics.

In the end, the only limit to this new species may be the availability of helium. The U.S. Department of the Interior reports that demand is outstripping supply, and shortages of high-grade helium have occurred. But standard-grade helium is still easily found at party stores, florists, and elsewhere — meaning there's plenty to float a fleet of robotic blimps in your living room.

➕ Jed Berk's Blubber Bots: degree119.com

Mark Allen is the founding director of Machine Project, a Los Angeles art and event space dedicated to exploring the connections between art, science, technology, music, and literature. machineproject.com

Photograph of children by Theo Alexopoulos

THE SAFE WORKSHOP

Rules to make by.

By William Gurstelle

Your workshop should be a welcoming and friendly place. The key lies in creating a safe and secure environment. Before embarking on a new project, it's a good idea to take a close look at the working conditions in your shop.

If your project area gives you a vaguely nervous feeling, then now's the time to bring things up to date. Don't delay — inspect, review, and evaluate your space and make whatever changes seem necessary to keep you out of trouble.

Don't know where to start? Here are some ideas from the members of MAKE's Technical Advisory Board to get you started. Have at it!

Wait 12 hours between sketching the plans and starting the construction process. The times people get hurt are usually when they're excited and in a hurry. Slow down, and work deliberately.

The high-decibel noise generated by power tools such as table saws and circular saws can damage your hearing. Protect your ears by using full-sized, earmuff-style protectors.

Wear a particle mask when appropriate to avoid breathing dust and other particulate pollutants that are common in workshops. Sawdust from treated wood and some plastics has known health risks.

Secure your work when using hand or power tools.

Avoid using a table saw when you can. Statistically, it's easily the most dangerous piece of equipment in the shop.

Obtain a pair of well-fitting, cool polycarb goggles, leather work gloves, and a protective lab coat. Make them attractive and stylish so that wearing safety equipment is fun. Pull back long hair.

Aim away from yourself. When cutting with a utility knife, position yourself so that when you slip, the blade doesn't land in your flesh.

Always use clamps, not your hands, to hold a work piece on a drill press table. If the tool binds, the work will spin dangerously.

Don't touch a bare wire, or cut any wire, until you're sure where the other end goes. When in doubt, measure the potential. This will save you from a possible heart-stopping electrical shock.

Always keep a first aid kit in your workshop, and always know where it is. First aid kits can be purchased ready-made, or you can put them together yourself. Essential items include bandages, pads, gauze, scissors, tweezers, and tape.

If you work with heavy things — say, timbers or an angle iron — or are prone to dropping tools, steel-toed safety shoes are a great investment in long-term foot appearance.

Install a smoke detector in your shop and place a fire extinguisher in an easy-to-reach spot. Make sure the extinguisher is rated for all types of fires.

Photography by Jason Madara

MAKER'S CORNER By Dan Woods, Associate Publisher

Tips and news for MAKE readers.

Maker Survey Completed

We just completed our most comprehensive survey of MAKE readers ever: more than 8,000 respondents worldwide! A big thank you to those of you who participated. We're analyzing the responses and will publish the results soon. If you'd like to put your name on a list for future MAKE-related research, including surveys and focus groups, send me a note at makesurveys@makezine.com. Just include your name and ZIP code in the body of the email. Everyone on the list will receive an advance copy of survey results when they're ready.

Boxed Sets Are Here!
MAKE: 3rd Year and CRAFT: 1st Year

The most-sought item in the Maker Store the past two years has been our MAKE collector's sets. In fact, only a few sets of MAKE: The First Year and MAKE: The Next Year remain. Each boxed set includes an entire year of MAKE or CRAFT packaged in a very cool collector's box — perfect gifts for makers who may have missed a few of our earlier issues, or an awesome gift idea for your local school science library. And if you'd like to purchase just the empty boxes to protect those loose volumes you've been saving, we carry those as well. Order yours today at the Maker Store, store.makezine.com.

MAKE Digital Edition

Every MAKE project and article ever published is accessible in a searchable online archive. Remember, MAKE subscribers can search the entire three-year collection of MAKE magazine, print workshop copies of projects, and share any article or project with friends. MAKE Digital Edition is absolutely free to MAKE subscribers. To subscribe, be sure to visit makezine.com/subscribe today.

Subscription or Delivery Problems?

If you have a problem with your subscription or magazine delivery (perish the thought), start with the customer support folks at cs@readerservices.makezine.com or (866) 289-8847 (U.S.+Canada) or (818) 487-2037 (all other countries). But if you're left less than satisfied, don't stew. Contact me at dan@oreilly.com and let me know about it.

The Holiday Maker Store

It's like The Little Engine That Could for makers. We've been scouring the planet to find unusual gifts for tech enthusiasts, teachers, backyard scientists, renegade crafters, hackers, students, inventors, and everyone who loves to do DIY projects in their backyards, basements, and garages. The Maker Store offers life-enriching challenges and exploration through DIY science, tech, and crafting projects for people who like to make stuff. We're not your gleaming, category-killer, big-box retail store, but if you're looking for the real deal for the makers on your list, stop by the Maker Store at store.makezine.com.

Here's just a small sampling of what you'll find:

The Designing Automata Kit — Learn about simple mechanics using cams and a crank slider mechanism. Many designs can be made, and because no glues are used, the kit can be used over and over again. We're the only store this side of the pond to carry this kit, made of chemical-free rubberwood from sustainable sources.

Arduino Diecimila Plus USB Board — An inexpensive tool for making computers that can sense and control the physical world beyond your desktop. An open source physical computing platform based on a simple microcontroller board, the Diecimila Plus includes a development environment for writing software for the board.

MAKE:it Electronic Maker's Toolkit — Handpicked by MAKE Senior Editor Phil Torrone, this kit includes everything you need to get started with a broad range of soldering and electronic projects featured in the pages of MAKE.

MAKE Controller — Born in the fire and smoke of Survival Research Labs' fabled Potrero Hill compound in San Francisco, the MAKE Controller comes ready to connect sensors, motors, servos, or maybe even your own killer (or peacemaking) robot.

MiniPOV2 — Never soldered? Don't know where to start? MIT engineer and MAKE contributor Limor Fried has cooked up this easy-to-build persistence-of-vision (POV) demonstration to show how microcontrollers work.

K450 PVC Rocket Engine Design & Construction Book — Detailed instructions on how to build a powerful K450 engine that will send a rocket soaring over 5,000 feet high!

Dan Woods is associate publisher of MAKE and CRAFT magazines. When he's not working on circulation and marketing or finding cool new stuff for the Maker Store, he likes to hack and build barbecues, smokers, and outdoor grills.

ART WORK
How Low Can You Go?

By Douglas Repetto

GERMAN ARTIST WOLFGANG LAIB LIVES in a remote village in the Black Forest. Each spring and summer he wanders the fields and forests near his home, patiently collecting pollen from dandelions, hazelnut trees, and other local flora. He uses the deep orange and yellow grains, stored in glass jars, to create powerfully simple, elegant artworks.

The Five Mountains Not to Climb On is a row of five small piles of hazelnut pollen sitting on the floor. Pollen from Dandelion is an enormous glowing square of, well, dandelion pollen. Laib's works are about as low-tech and DIY as possible; he hand-gathers ubiquitous, but seldom seen (if often sensed!), materials in enormous quantities, and then presents them in unexpected contexts.

British artist Andy Goldsworthy works with similarly simple, unprocessed materials, although his creations are even more labor intensive. He often creates temporary, site-specific works from color-sorted leaves, stacked and ordered rocks, masses of twigs, and so on. While Laib collects materials and transports them to new contexts, Goldsworthy typically reorganizes them in place. (The documentary Rivers and Tides is a great introduction to his work.)

The force of these works derives largely from the friction between their simplicity and their striking physical presence. Laib's pollen is luminous; it smells good, and even in very small quantities it seems like a mysterious, precious substance.

Goldsworthy's materials often seem highly unlikely, out of place, yet clearly they're literally of the place. There's something subtly disconcerting, but thrilling, about seeing a rock completely covered with a smooth gradient of color-sorted leaves, like a physical Photoshop filter created and applied entirely by hand.

Aside from their physical strangeness, there's another aspect of these works that makes them even more compelling to me: Laib and Goldsworthy seem to have reached a kind of DIY nirvana. They use nothing purchased, nothing manufactured, nothing shipped from halfway around the globe.

They literally walk into an environment, collect their materials, and get to work. Both have various conceptual and aesthetic reasons for working as they do (Laib is broadly concerned with spiritual/ritualistic questions, Goldsworthy with environmental issues), but in practical terms they've spent their lives taking the DIY ethos to its limits.

There is something about the DIY spirit that often leads to a wonderful kind of infinite descent into extremism. There's always a way to take things a bit further, to cut out another middleman, to make things just that much more difficult for yourself, but also that much more fun. Some people spend months hand-collecting pollen, others make their own metals by smelting ores in a homemade furnace, or spin their own yarn from tumbleweeds and dog fur.

Doing things entirely yourself isn't a negation of technology, or some sort of Luddite accusation; it's more of a thought experiment made real. It can help clarify what, exactly, the technologies we're using are good for (and many of them are, of course, very good), or when they're simply wasting resources and adding unnecessary complexity or cost.

Now, to someone trying to build, say, a self-replicating 3D printer, collecting grains of pollen or stacking some twigs and then calling it a day might not seem like such a glorious DIY achievement. (And I bet Laib doesn't even fire his own glass jars, the lazy sod!) But don't underestimate the difficulty you can face, nor the satisfaction you can feel, when working at an extremely base level with manual techniques and simple materials.

MAKE mycology maestro Philip Ross has recently

been teaching art classes that focus on super-basic skills like making stone tools, fire starting, salt harvesting, and so on. These aren't exactly typical topics for an art class, and while most of his students probably won't end up using the skills directly in their creations, the experience of learning how to do such things, and the realization that they are often exceedingly difficult, will certainly help them approach their work in new ways.

Artist/inventor Natalie Jeremijenko's students are following a similar thread, but in reverse. On howstuffismade.org, students attempt to trace the component parts of a manufactured product back to their raw material origins. For some products, the path is simple: wine giants Ernest and Julio Gallo make their own bottles directly from silica, limestone, and soda ash. Other paths are much more complex, and even nearly untraceable.

Check out Michael Pollan's recent book *The Omnivore's Dilemma* for a food-oriented take on the surprisingly complex and morally ambiguous task of tracking the sources of four different meals (fast food, Whole Foods, boutique organic, modern hunter-gatherer).

Yet even the most complex technologies are ultimately reducible to their component sources. We don't (yet) have alien technologies that just fell off the back of the spaceship; in fact, NASA has some great videos of spacecraft parts being laboriously hand-machined. Singularity-induced, hyper-nano-cyber manufacturing has not yet made human ingenuity redundant, although hyper-global-miniaturized-CAD-driven manufacturing has certainly made reverse engineering a lot more difficult than it used to be.

It's clear that we can't all be full-time hunter-gatherers, or home-smelters, or dog fur spinners. And there are good environmental and social arguments for not doing some things yourself. (Please don't start a backyard mercury mine!) There's no going back to a mythical pure state where we're all self-sufficient, environmentally pure master makers. But we can search out interesting points on the paths between high-tech and no-tech, mass-produced and handmade.

There's always another way to do things, and there's great joy (and sometimes, great virtue) to be had in exploring complex ideas via simple means. It'd be pretty boring if all art were made from pollen and twigs, but that doesn't mean that "simple is good" and "less is more" aren't still powerful ideas. So, a challenge, a thought experiment, waiting to be made real: how low can you go?

Laib and Goldsworthy seem to have reached a kind of DIY nirvana. They use nothing purchased, nothing manufactured, nothing shipped from halfway around the globe.

TEMPORAL ARTWORKS BY ANDY GOLDSWORTHY: *Sumach leaves laid around a hole* (top), and *Leaves laid over branch ring laid on a rock* (bottom).

Douglas Irving Repetto is an artist and teacher involved in a number of art/community groups including Dorkbot, ArtBots, Organizm, and Music-dsp.

Photography courtesy of Galerie Lelong, New York

THE ITCH TO ETCH

By Tom Owad

THERE ARE MANY WAYS TO DO INTRICATE metalwork, but most of them involve expensive machinery or high-level training. One method, though, is accessible to all: chemical etching. This method is very similar to that used to etch hobbyist circuit boards: a mask is placed on the metal, and the non-masked area is dissolved by chemicals.

Brass is well suited to the process and produces attractive results. Shapiro Supply will sell you a .078" thick, 8" square piece for $11. See their listings on eBay (myworld.ebay.com/ssshapiro).

Design your mask and print it out on a laser printer using inkjet paper. Keep in mind that the mask covers the areas that do *not* get removed. The project shown here is a Victorian-style panel for a 20×2 LCD. You can find this design, and others by Andrew Lewis, in the gallery section of monkeysailor.co.uk.

Scrub the brass clean and then wipe with iso-propyl alcohol. Place your print facedown on the brass and iron it for a minute or two, until the toner heats and sticks to the brass. Now soak the plate in a dish of water for 10 to 15 minutes, or until the paper is sufficiently soft that you can remove it. Remove the paper, then mask the back of the brass plate by covering it with gaffer's tape (like duct tape, but it doesn't leave so much residue).

Next, prepare a solution of ferric chloride ($FeCl_3$). It's available as a liquid solution from E-Clec-Tech (e-clec-tech.com) or as a solid you have to mix yourself from United Nuclear (unitednuclear.com).

⚠ **WARNING: Your ferric chloride should come with a Material Safety Data Sheet. Wear goggles, as it will blind you if it gets in your eyes. Also wear gloves and work in a well-ventilated area. The fumes are extremely noxious. Anything the ferric chloride touches will be permanently stained orange. More safety info is available at makezine.com/go/fecl3.**

If you're looking on eBay, you'll notice you can get *anhydrous* ferric chloride quite cheap. Avoid this, as it gets extremely hot when mixed into a solution and must be added to the water very gradually. If you add water directly to the anhydrous crystals, it has the potential to flash into steam, spraying your face with boiling hot $FeCl_3$. And you'd have to buy more ferric chloride, eliminating the cost savings.

ETCHPUNK: Brass cover for LCD display.

Dip the plate into the ferric chloride and let it sit for 10 to 30 minutes. The time required depends on the temperature (warmer is faster) and the depth of etching required. When you're satisfied with the etch, remove the plate and rinse it in water. Wipe clean with wire wool.

That's it — a bit of ferric chloride and a laser printer are really all you need. This particular project requires some drilling and cutting to finish the LCD panel, but if you're just doing something decorative, you're done.

Copper can be used as easily as brass. Aluminum can also work, but be careful, as it gets very hot and gives off noxious fumes. Use a weaker solution.

It's also possible to cut all the way through the metal. Suppose you'd like to make a cog. Print out 2 copies of the cog, and iron one to each side of the plate, lining them up. The ferric chloride will eat around the cog's mask on both sides, leaving a hole where the 2 sides meet in the middle, and leaving the shape of the cog intact.

Tom Owad is the owner of Schnitz Technology, a Macintosh consultancy in York, Pa. He spends his days tinkering and learning, and is the owner and webmaster of applefritter.com.

Photograph by Tom Owad

COOL GIFTS *for smart people* *who like to*

» tech enthusiasts, teachers, backyard scientists, renegade crafters, hackers, students, and backyard inventors

SPECIALLY SELECTED BY THE STAFF OF MAKE

MAKE STUFF

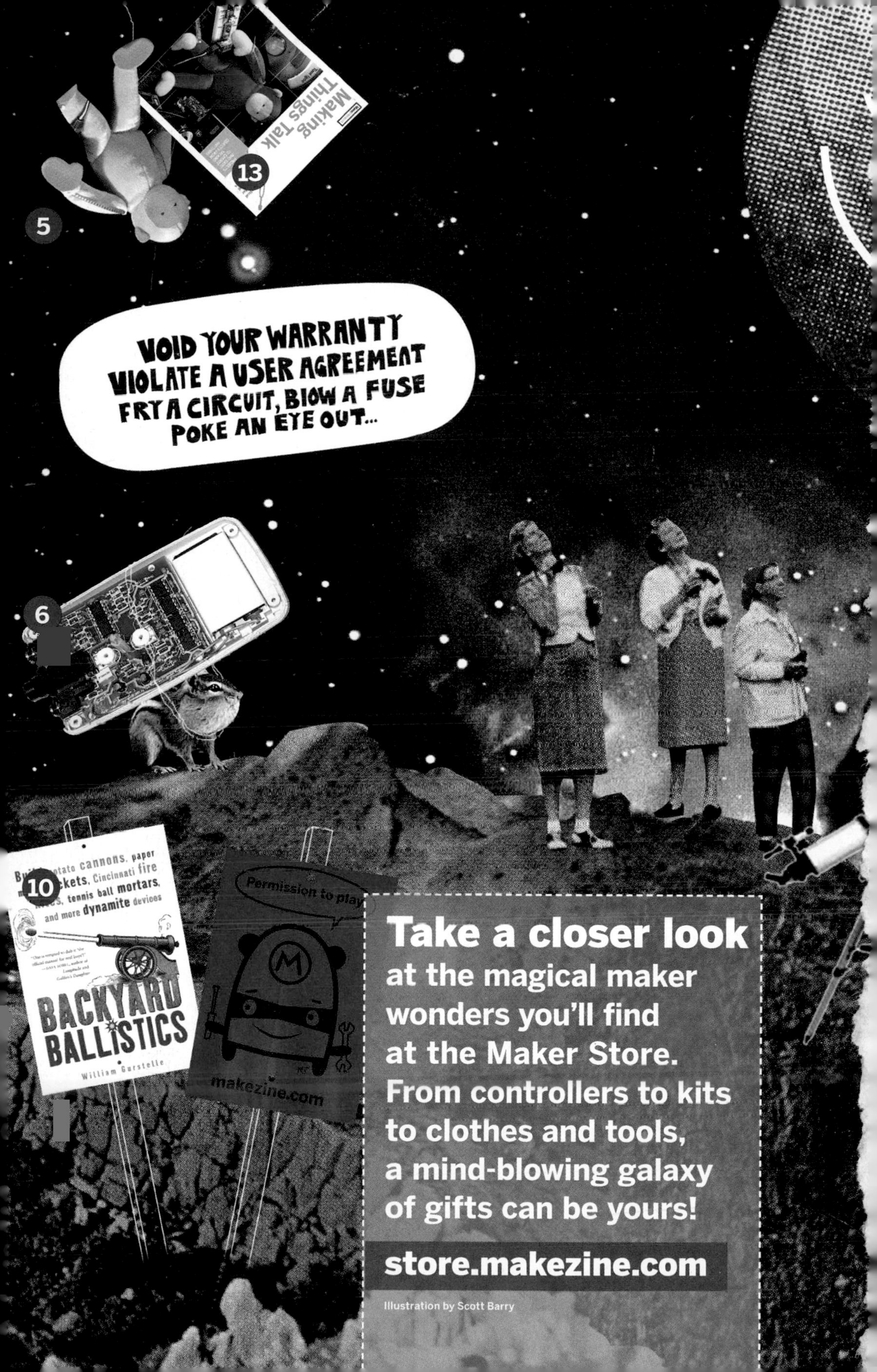
Making Things Talk
13
5
VOID YOUR WARRANTY
VIOLATE A USER AGREEMENT
FRY A CIRCUIT, BLOW A FUSE
POKE AN EYE OUT...
6
10
BACKYARD BALLISTICS
William Gurstelle
Permission to play
makezine.com
Take a closer look
at the magical maker wonders you'll find at the Maker Store. From controllers to kits to clothes and tools, a mind-blowing galaxy of gifts can be yours!
store.makezine.com

Illustration by Scott Barry

A world of wonders awaits you at the Maker Store,
open 24 hours a day at **store.makezine.com**.

MAKE BOX SETS

Two years ago a few of us were sitting around the kitchen joking about how we ought to take a lead from *The Sopranos*: The Complete First Season and create MAKE: The First Year. So we did. And then MAKE: The Next Year. And now ... you guessed it: MAKE: The Third Year. Thousands of these puppies have been snatched up by makers from around the planet who need (and we stress *need*) to know what they missed. Plus, the collector's box is built to outlast most makers. Can you say "long tail"? You want an awesome holiday gift? This is it. **$59.99**

and subscriptions!

SAVE OVER 10%

BUY A GIFT
SUBSCRIPTION TODAY
(OR 1 FOR YOURSELF)
FOR ONLY **$31**

Go to:
makezine.com/gift

USE PROMO CODE
47GIFT

« SEE INSIDE

KITS

① BLUBBER BOTS $99.99
Blubber Bots are DIY robotic inflatables that navigate autonomously and intelligently. These helium-filled, light-seeking balloons graze the landscape in search of light and cellphone signals. Launch one or launch a herd.

② MAKE CONTROLLER $149.95
Born of Survival Research Labs' fabled compound in San Francisco, the MAKE Controller comes ready to connect sensors, motors, servos, and maybe even your own killer (or peace-making) robot. This next generation of modular, programmable controller boards is the brainchild of MAKE magazine and MakingThings.

③ MINTY BOOST $19.99
Everything you need to make a small and simple (but very powerful) USB charger for your MP3 player, camera, cellphone, and any other gadget you can plug into a USB port to charge! Even if you've never soldered before, this is a relatively straightforward project kit.

④ MINI POV $17.99
Never soldered but want to? MIT engineer and MAKE contributor Limor Fried cooked up this easy-to-build persistence-of-vision (POV) demonstration to show how microcontrollers work. Perfect for students of all ages.

⑤ ARDUINO DIECIMILA $34.99
An inexpensive tool for making computers that can sense and control the physical world beyond your desktop computer. This open source physical computing platform, based on a simple microcontroller board, includes a development environment for writing software for the board. (Monkey not included.)

⑥ HIGH SPEED PHOTO KIT $120
Take "impossible" pictures that will amaze your friends. Capture high-speed events: a splash, popping balloons, breaking glass. This kit includes a disposable camera with high-speed flash and an adjustable flash controller with a fully assembled flash trigger that synchronizes the high-speed event and the flash.

TOOLS & APPAREL

⑦ MAKE: WARRANTY VOIDER $39.95
Small enough to fit on your key chain, this Leatherman "Squirt" P4 is the perfect companion for mobile fixing, hacking, and MacGyvering.

⑧ PERMISSION TO PLAY KID'S SHIRT $10
Inspired by a heartwarming letter from a reader thanking us for giving him "permission to play," this shirt has been flying off the shelves since it debuted at Maker Faire last spring.

⑨ HOODIES $21.99
Wear the open source hardware mantra with pride: "If You Can't Open It, You Don't Own It." Stay warm and start a worthy conversation with a total stranger. We sell a truckload of these in the first couple hours of Maker Faire Bay Area before the fog burns off.

BOOKS

⑩ BACKYARD BALLISTICS $16.95
Projects ranging from a match-powered rocket to a tabletop catapult to a tennis ball cannon. C'mon. What self-respecting maker doesn't need this book? The title alone makes it worthy of a prized place in your library.

⑪ ILLUSTRATED GUIDE TO ASTRONOMICAL WONDERS $29.99
With the advent of inexpensive, high-powered telescopes, amateur astronomy is fully within reach, and this is the ideal book to get you started. Offers you a guide to the equipment you need, and shows you how and where to find hundreds of spectacular objects in the deep sky.

⑫ BEST OF MAKE $34.99
Need we say more? We thought not!

⑬ MAKING THINGS TALK $29.99
Tom Igoe bestows the power of communication upon your favorite tech creations through simple projects that present the guidelines for electronic verbosity. Whether it's microcontroller-powered devices, email programs, or networked databases, Igoe demonstrates the ability of electronics to interact in fun and interesting ways.

re-use
re-cycle
re-Make

4

PATENT-B-GONE!

Inventor Mitch Altman explains why he open-sourced his TV-B-Gone kit.

As an inventor, I was taught that patents encouraged creativity and entrepreneurship. So, after finishing my first TV-B-Gone universal remote control prototype, I naturally called my brother the patent attorney, and together we filed a patent application.

Was that the best move?

TV-B-Gone remote controls are key chains with one button that make it fun to turn off almost any TV in public places. Oddly enough, within weeks of the first day of sales, the TV-B-Gone story appeared in major and minor newspaper, magazine, radio, and even TV outlets throughout the planet. It was a hit!

With this vast popularity, what might have happened if my packaging had not displayed the words: "Patent Pending"? Maybe it stopped some large companies from copying TV-B-Gone remotes, since selling copies would open them up for being sued once my patent was granted.

Would it be different if my product were open source?

I knew about open source, of course, but never considered it viable for hardware until going to my first hacker convention. There I met people who are very critical of patents and other forms of intellectual property law. They see these laws as obsolete and obnoxious. Individuals who want to hack cool ideas to improve upon them and share their results are often preyed upon and silenced by corporate lawyers protecting their clients' patents. Paradoxically, this stifles the creativity that patents were supposed to encourage. This point of view was an eye-opener for me.

Even though my project was not open source, I benefited from the open source community. People hacked TV-B-Gone remote controls in wonderfully creative ways. (Search online for "TV-B-Gone hacks" and you'll get the idea.) These hacks increased the product's popularity, resulting in more sales and more people around the world experiencing the satisfaction of turning off TVs. Also, since there was an army of TV-B-Goners who emailed me with ideas on how to improve upon

my initial design, the next versions of TV-B-Gone remotes were considerably better than the original. Everything added up for me to look seriously at Creative Commons, a form of open source licensing.

I decided to go for it. Together with Limor Fried (who makes lots of great kits), we're making open source kits available so anyone can build and hack TV-B-Gone remote controls (look for an upcoming MAKE article about this). The firmware source code will be available online, as well as the board layout, lots of TV power codes, and all documentation.

The added buzz will likely also help sales of ready-made TV-B-Gone key chains, since not everyone wants to build their own. Everybody wins. In the words of my brother the patent lawyer, "The old way of patent law is to think: 'This is mine and I'm going to keep it.' This may have some advantages, but with open source you can share and bring more creative minds to the process. What's really nice is that you don't have to give up all your rights. With open source you can have the best of all worlds."

Mitch Altman's next products, based on the "Brain Machine" article he wrote for MAKE, Volume 10, are also open source.

UPLOAD

Opening the door to digital arts and crafts.
By Charles Platt

The projects in MAKE encourage us to transcend our default role as passive consumers. Armed with screwdrivers and soldering irons, we boldly go into basement workshops, creating new gadgets or ripping open old ones, sometimes achieving mixed results but always enjoying ourselves.

Earlier this year, it occurred to me that the magazine could extend its interests from the physical world into an area that I think of as "digital arts and crafts": photographs, videos, music, text, computer code, and animations. The snag is that the software involved is increasingly diverse and complex. Almost anyone knows how to use a hammer, but how many of us have the time and patience to enhance a video with Adobe After Effects — and retrain ourselves each time an upgrade is published?

Numerous magazines are dedicated to specialty tasks such as photo retouching or sound synthesis, but what I want is a broader view of the whole digital-arts spectrum, featuring small-scale, specific projects that will be fun, quick, and easy to complete.

Because I was unable to find such an overview, I was excited by the opportunity to assemble it in this new section. Under the broad title of Upload (meaning anything digital that can be uploaded via email or to web pages) you'll find projects ranging from chroma key video to infrared photography to online book publishing. In the future I hope this section continues on a regular basis — but this, of course, will depend on you. Do you have a new and clever application of a digital tool, to achieve an unexpectedly creative product? Be sure to let me know.

—*Charles Platt, Upload Section Editor*
platt@makezine.com

Looking at the Low End

📷 Infrared photography reveals a world invisible to the naked eye.
By Richard Kadrey

For the human eye, the lowest visible wavelengths are red light measuring about 700 nanometers (nm). Below that, infrared radiation runs from about 750nm down to 1mm. When photographed in this part of the spectrum, leaves and grass glow with energy, as if the entire natural world is lined with fiber optics. Skin is luminous and perfect, like alabaster. Infrared photography gives you an inhuman view of the world, and it's a beautiful one.

In the beginning, infrared photography was nothing you needed to know about. It was a high-tech procedure reserved for laboratories and mapping satellites. Even when artists got their hands on the stuff, it required special film that had to be kept in an ice chest until it was used, and special processing that required access to a darkroom with the right chemicals, and all the expenses those items entailed.

Digital photography has made infrared accessible to everyone. That's great news to those using IR for the first time, because this is when you're liable to make the most mistakes. Better yet, you don't need an expensive camera to take great shots. In fact, cheaper and so-called "dinosaur" digital cameras can be the best ones for IR shooting. The reason is simple: most high-end cameras come with a built-in infrared-blocking filter (sometimes called "hot glass") that sits right in front of the camera's sensor chip. Cheap cameras don't always have this IR filter, and they're easy to hack if they do. But remember when picking your cheap camera to make sure it has a Preview mode. This will allow you to see your infrared shot and make adjustments on the fly.

Photography by Richard Kadrey

1. GET A DIGITAL CAMERA THAT CAN SEE INFRARED

You may already own a camera that can see infrared. To find out, point a TV remote control toward your camera (in low room lighting), press any button on the remote, and watch it on your camera's LCD screen. If you see the tiny lens on the end of the remote glowing white, your camera is sensitive beyond the human visible spectrum.

I shoot infrared with a Sony Cybershot DSC-F707. When it was released in 2001 it was considered a fairly high-end "prosumer" camera. Most photo geeks now think of this 5-megapixel machine as a kind of steam-powered relic, something that would have wowed Jules Verne, but that's just amateur-hour snobbery spouted by people who think that the only thing that counts is who has the bigger megapixel count. Even shooting in regular mode, it's not hard to make good 8×10 prints with the F707. In fact, 95% of the prints in my first gallery show were

shot with the F707. And it packs plenty of power for infrared work.

One reason the Sony F707 is such a great off-the-shelf infrared shooter is that it comes with a NightShot feature, which is a built-in infrared system. Unfortunately, this works only in the dark.

Sony crippled the ability to shoot IR in daylight because, supposedly, some fabrics are transparent to infrared, effectively turning all of Sony's cameras from that period into a voyeur's favorite new toy. But any hack a corporation comes up with can be broken by patient geeks on a mission. By putting the F707 in NightShot mode and adding infrared and neutral density filters, we trick the camera into thinking that daytime is nighttime.

Infrared photo wiz Chris Maher shoots with other dinosaur cameras, such as the Nikon 950 and 990. I loved the 950 and 990, generally loathed by "real" photographers when they came out. Their bodies swiveled independently of the lens, so using

Preview mode, you could shoot over and around people without anyone knowing what you were doing.

Another favorite of infrared enthusiasts is the old Olympus 3030. Instead of trying to unload it for $10 in a garage sale, you can use it to see the world in a whole new light — literally.

2. GET SOME INFRARED FILTERS

By hacking your digital camera, you're tricking it into thinking that it's still shooting in the visible spectrum. To do this, all you need are a few simple filters. When you put the filters on your lens, you can forget about all your regular camera settings. This is why you need Preview mode.

The rim of the F707's lens, like most lenses, is threaded so that you can screw on standard filters, such as polarizers for outdoor shooting. If your lens doesn't have threads, you can simply hold a filter over the lens with one hand and shoot with the other.

Infrared filters come in all the standard lens sizes. The F707 takes 58mm filters. The first IR filter I used was the Hoya R-72, which allows only infrared rays longer than 720nm to pass through. Other infrared filters block different parts of the spectrum, giving your shots different looks, bringing out different details in the sky, foliage, and foreground objects.

The best way to find out what works for you, and works in different situations, is to experiment. Since you can pick up an F707 on eBay for $225–$250, you may have some money left for filter shopping. Filters can run from $35–$200, depending on the size of your lens.

3. GET SOME NEUTRAL DENSITY FILTERS

Along with your infrared filter, you may need 1 or 2 neutral density filters to cut the amount of visible light entering your camera, without changing its color. Think of it as a piece of welding glass you could use to look at a solar eclipse. My first neutral density was an ND-400.

Without an ND filter your infrared shots can be much too bright. On a sunny day, the highlights will be completely blown out. Even with a single neutral density, you can end up with too much light. Since the filters come threaded, you can add a second ND, along with the infrared filter.

THE LOOK OF INFRARED: Infrared radiation reflects off solid surfaces at different frequencies, depending on the surface's composition. Leaves and branches can turn snowy white, and water an opaque black.

4. LEARN MORE

One of the best sites on infrared shooting is infrareddreams.com, with an excellent IR primer at infrareddreams.com/how_to_shoot_ir.htm.

For a more rigorous and technical exploration of infrared photography, with many camera comparisons, visit dpfwiw.com/ir.htm.

You can find infrared filters at most decent camera shops, as well as good online photo sites such as bhphotovideo.com. You can even find them in Amazon's Camera area.

Have fun shooting. Experiment. Take chances. And be prepared to see a world that you have never seen before.

Richard Kadrey has written about technology and culture for magazines such as *Wired*. He's also a fiction writer; his latest novel is *Butcher Bird*.

Book Yourself

Innovative options enable you to publish your own text and pictures.
By Kevin Kelly

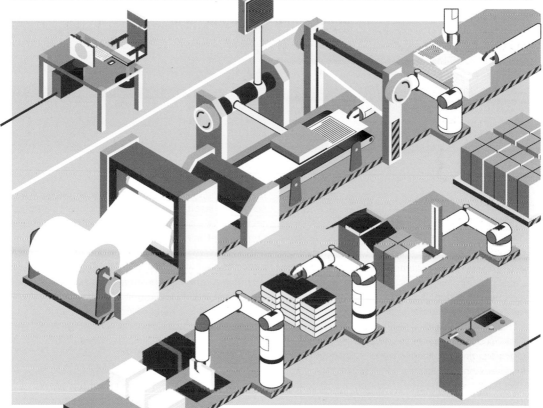

Approximately every minute, someone in the world publishes a new book. We are in a golden era of publishing and broadcasting. These days the mechanics of making a book are almost as trivial and as easy as starting a blog.

While mainstream New York mass-market publishers struggle to sustain their traditional system of distribution through bookstores, the choices for self-publishers continue to expand and mature.

You can now create a book in paper, in hardcover, in color or black-and-white, as a downloadable PDF, or as an e-book. Smart self-publishers will try to exploit all these options.

The two basic ways to manufacture a book are to print a bunch, or print them one by one on demand.

The advantage of printing on demand is that you print a copy only after you sell it. This eliminates the costs of storage and of unsold books, and also self-finances the printing.

However, while the cost of print on demand has fallen significantly, it's (and likely always will be) cheaper per unit to print up a bunch of books. And sometimes, you really do want a bunch of copies — for a conference, a book club, Christmas presents, or for a best seller.

ON-DEMAND PHOTO BOOKS: Lulu.com and blurb.com produce beautiful color books on demand (above). A large book of photos taken by the author on several trips to Asia (upper right). A mini book of Burning Man photos taken by the author (lower right).

METHOD 1:
BATCH PRINTING

Luckily there's an accessible technology for this. Xerox's DocuTech system, leased by many copy centers around the country, is able to print multiple copies of a title fairly quickly and cheaply. When used with auxiliary binding equipment, it can produce short-run books.

This system is really only an option for standard black ink on white paper, in standard trade paperback sizes, and you need to start with a completely designed book, preferably in PDF format.

For a print run of, say, a minimum of 250 copies of a 200-page 6"×9" book (with color cover), you can probably find someone to print it for around $3.50 per book. The quality will be very close to, if not indistinguishable from, a store-bought softcover book. The disadvantages of this process are that you'll have to invest the cost of printing all the books upfront (in this case $875), pay the shipping if the printing is not local, and then store boxes of books.

This is a fast-moving field, with new technology all the time, so the best printers for a particular job change very quickly. A few short-run printers that I've used with satisfaction in the past are DeHart's (deharts.com) and Commercial Communications Inc. (comcom.com).

If you print your own book, you'll want to offer it for sale on Amazon, even if you'll be selling it directly from your website. You need 2 copies of the book to send to Amazon, and you need to include a barcode on the book's back cover before you print it. The steps to getting your book listed on Amazon are not obvious, so I've written up a separate tutorial at kk.org/cooltools/archives/000668.php.

Having a book listed on Amazon does not preclude any other options you might want to use, but the costs of keeping it there reduce its profitability, so it should not be your only option.

METHOD 2:
PRINT ON DEMAND

The most exciting frontier of self-publishing books today is print on demand, usually employing toner-based or inkjet-like technology to print single copies. You send a digital file to the printer, who runs off individual copies of the book. You can store a digital file of your book at the printer, and then your customer simply orders the book at a price you set, using a credit card or PayPal. It's then shipped to them, and you keep whatever profit you have assigned yourself.

Three web-based companies provide comprehensive on-demand services for self-publishers: iUniverse, Virtualbookworm, and WingSpan Press. Each of them offers package deals for $300 to $500. For this you get basic editorial help and get your book designed and proofed, with a nice custom

Photography by Kevin Kelly

cover, a barcode, a listing on their online store, and a few copies for yourself. Design-clueless authors may find this attractive.

But anyone with even a mild sense of what looks OK can manage a better way. Produce your book using a layout program that can export a PDF file. Proof and then upload the PDF to lulu.com. Lulu will print your standard black-and-white softcover book for about $8 per book. You can ship books to yourself, or set a price above your $8 cost and let readers order the book directly from the Lulu website.

Printing a book in full color is expensive. A regular 6"×9" softcover book in full color throughout would run about $30 on Lulu. A better deal for color printing, and for photographers and artists in particular, is blurb.com. Blurb uses a format similar to Apple's iPhoto books to produce exquisite coffee table books, but at about ⅓ the cost of Apple. A softcover 8"×10" photo book with 80 color pages runs about $30.

I've been most excited about Blurb's recent option of a huge 11"×13" portfolio book. I made several 120-page books this size, containing some 500 photographs. They are simply awesome, almost overwhelming, with better quality printing (using HP's Indigo press) than most art books. In copies of one they cost $70 each. Blurb's only downside is that you have to use their design templates, unless you want to do some fancy exporting of files. But it sure is easy. I now make a new book per month. They are ideal as gifts recounting a visit or trip, or as mini-portfolios and scrapbooks.

It's fairly simple to make digital books. The easiest way is to render your file — text or images or both — as a PDF. You'll want to optimize it so that it's as compact as possible. Be sure to include the cover. To read it, all anyone needs is Acrobat Reader on a PC. You can also sell the PDF version of the book as an e-book. Lulu and the other self-publishing clearinghouses offer options for selling e-books.

If you want to sell e-books from your own website, you can sign up for an account with payloadz.com, which delivers a digital copy after someone pays for it using PayPal.

I recently experimented with offering my self-published books in every possible format, while setting my prices so that I made exactly the same profit from each channel. I decided I'd be happy to make $1.50 per book, no matter whether I sold a bulk-printed $9 copy via Amazon, or an on-demand color copy for $27 from Lulu, or a $2 PDF version from Payloadz. In each case I made the same $1.50 after I deducted my total costs, so I was totally agnostic toward the format and had no incentive to push one or the other. Let the customer decide! The results? I sold ten times as many digital download PDF versions of the book as printed versions. It's really the future of publishing.

Kevin Kelly is Senior Maverick at *Wired*, and publisher of Cool Tools, True Films, and Street Use blogs (kk.org).

Seeing Red

Shifting the spectrum can transform a landscape and create dramatic artistic effects. By Charles Platt

Back in the day when monochrome prints dominated art photography, big names such as Ansel Adams created dramatic effects by using colored filters with black-and-white film. A red filter, in particular, blocks light from the blue sky while freely transmitting the mellow colors of rocks and dry grass. This combination results in an almost black sky while the bleached foreground seems to leap out at the viewer.

Today we can achieve the same results much more easily with Photoshop. The software is expensive, but older versions still work. The images in this how-to were created with version 6, available legitimately for around $50 on eBay. Newer versions make it easier to simulate filtering, but doing it the old way will give you a clearer idea of how it works, and will put you in a better position to control the process.

1. ACCENTUATE THE COLORS

Begin with a photograph that has a rich range of colors. From the Menu bar in Photoshop choose Image ⇒ Adjust ⇒ Hue, Saturation and push the Saturation slider to +10 or +15.

2. CREATE A RED LAYER

From the Menu bar, choose Layer ⇒ New Fill Layer ⇒ Solid Color and in the dialog box that opens, pull down the Mode menu and choose the Multiply option (to make the red filter transparent). Click OK, and then when the color picker window pops up, select maximum red. You can do this by looking for the data entry fields labeled R, G, and B. Enter 255 for R and 0 (zero) for G and B. Photoshop creates a red layer, showing you how Ansel Adams might have seen his subject through the viewfinder of his film camera. Remember, you're going to make everything monochrome in the next step. Bright red will become white, and mid-red will become gray.

3. CONVERT TO GRAYSCALE

From the Menu bar, choose Image ⇒ Mode ⇒ Grayscale, click Flatten in the Layers dialog box that appears, and you'll see the result: a black-and-white image.

4. TWEAK THE CONTRAST

The image needs more contrast, so, from the Menu bar, choose Image ⇒ Adjust ⇒ Auto Levels, and you're done.

5. WHAT IF YOU WANT A DIFFERENT COLOR FILTER?

Undo twice to get back to the layer you created during Step 2. If the Layers palette is not visible on your screen, choose Window ⇒ Show Layers from the Menu bar.

In the Layers palette, double-click the red thumbnail in the layer directly above the thumbnail of your photo. The color picker pops up. Now you can enter different values for R, G, and B, or click on a color in the picker window. This will be your new filter color.

Repeat Steps 3 and 4 above.

Photoshop 6 introduced a Channel Mixer option, which makes this whole process easier, although harder to visualize. Start with your color photo as before. Choose Image ⇒ Adjust ⇒ Channel Mixer from the Menu bar. Click the Monochrome checkbox at the bottom corner of the dialog box and move the sliders to and fro for instant black-and-white output, as if you were applying filters of any imaginable color.

Would Ansel Adams have approved? Probably not. If it had been this easy, everyone would have been doing it!

TOP LEFT: Near Edwards Air Force Base, Calif., a desert at sunset is classic Ansel Adams material. Colors have been pushed so that the subsequent steps will be optimized. TOP RIGHT: Impose a red filter. LEFT: Then convert to monochrome and adjust levels to get the Ansel Adams look.

LEFT: This picture was scanned from a 4×6 photo from a cheap processing lab that pushed the colors to the max.

CENTER: If you convert to mono-chrome mode in Photoshop (or in your scanner), the results are drab.

RIGHT: Using a red color filter, you get something much more interesting. (A speckled sky will result from the red filter discriminating between slightly different hues of blue.)

Go Green!

Special video effects are available to anyone with a cheap camcorder and $25 of software. Greenscreen is the most powerful of these, and is surprisingly easy to use. By Bill Barminski

Would you like to make a video of yourself standing on the moon? There are two ways to do it. You can build a rocket and fly there — expensive, not to mention dangerous. Or you can use a greenscreen to make it look as if you are there. Yes, a greenscreen. I hope I won't be shattering too many illusions when I tell you that this is how they did a lot of that cool stuff in *Star Wars*. They placed an actor in front of a greenscreen and filmed the scene while he pretended to fight a giant space squid. A technique called *chroma keying* was then used to remove the green color, allowing a new piece of video to be placed behind the actor.

This is called a *composite shot*, and the process is called *keying*. In the past you needed high-end software costing hundreds if not thousands of dollars, but today you can do it for $25 plus some cheap paint and lights.

But just let me issue a word of caution: green-screening can be tricky. There are many variables that can affect the outcome. Even professional filmmakers run into unexpected problems from time to time.

Photography by Bill Barminski

1. MAKE A BACKDROP

First you're going to need a screen, which can consist of colored fabric or a painted wall. Lime green is most commonly used, because it is so freaking ugly that the exact-same color is unlikely to appear on anyone or anything else in the shot, and thus it can be earmarked for replacement. (This means your subject can't wear a lime green tie.)

You can buy special greenscreen fabric and paint, but they're expensive. I've used very cheap green fabric from the local fabric store with decent results. I've even used a lime green blanket I found at a thrift store for $4. Look for something sheer that resists wrinkles, which will show up and make it harder to pull your key; iron the wrinkles out if you need to.

If you have a wall you can paint, so much the better, since there will be no wrinkles. Go to any paint store and pick out the worst lime green color you can find. Be sure it has a flat finish, not glossy. The exact shade is unimportant, since our software will find it for us when the time comes to replace it with our desired background image.

2. LIGHTING AND PLACEMENT

The biggest problems with greenscreen shots stem from poor lighting and placement of your subject. You want to illuminate your greenscreen with a flat, even light, so that it has no shadows or highlights. Don't use spotlights for this.

The placement of your subject in relation to the greenscreen is also crucial. The subject needs to be as far from the green as possible, to avoid picking up reflected green light. This is tricky because the reflected green is hard to see. Of course the farther away you put your subject, the bigger your green screen must be. If you're doing this for the first time, frame your subject from the waist up. Don't try a wide shot of the whole person.

The cheapest lighting source is the sun. If you can shoot outdoors, that's great, provided you find a place that gets even lighting with no shadows on the background. A gray, overcast day is actually best for shooting since it produces an even, flat light.

If you shoot indoors, you'll need 2 sets of lights, one for the greenscreen and one for the subject. Don't try to use the same set of lights for both.

To illuminate the greenscreen you can use cheap fluorescent tubes. They give a smooth, even light.

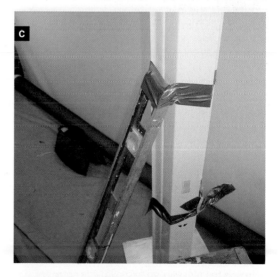

Fig. A: Fluorescent fixture and tubes ($12), scoop light ($10), camera (duh), duct tape, fake beard (optional), green blanket (about $4). **Fig. B:** Here's a greenscreen that's been thumbtacked to the wall. Be sure to stretch it tight to avoid wrinkles. **Fig. C:** This is why you need duct tape, unless you want to spend 200 bucks for light stands. **FACING PAGE:** The author in front of the "fire."

GREEN SCREEN TIPS. Fig. D: The subject has been filmed against a greenscreen. A mistake made here is the use of glass. As you can see, the green shows through. Avoid mistakes like this.

Fig. E: Subject after the green has been keyed out and new video inserted. The problem with the glass is evident. Other issues show up too, such as the green bounce light hitting the side of the chair. Color correction can help to remove or limit the green.

For about $12 you can buy a 4-foot fixture containing 2 tubes. Depending on the size of your greenscreen you may need 3 or 4 of these fixtures. I use sticks and lots of duct tape to anchor them. You can then light your subject with a couple of workshop clip lamps using bulbs rated from 100 to 500 watts. Remember that they must not cast shadows on your screen.

3. KEYING YOUR VIDEO

Now, let's say you've shot your footage and you're ready to key out the green. On a Mac, you can use iMovie with a plugin called Stupendous Software Masks & Compositing, which costs $25. If you have Windows, you can find free software such as ZS4 (zs4.net/downloads), or economical all-purpose editing software (with greenscreen feature included) such as Video Edit Magic, available for free in a trial version or for $69 fully featured, from trusted sites such as makezine.com/go/tucows.

3a. In iMovie, first import the video that you shot against a green background, and place it in the timeline. Then import the footage that will replace the background, and place it next to your video.

Select the first clip, go to the Effects category, and choose Green Screen, Smooth. This effect has 3 controls: Outside Fill, Inside Fill, and Choke. Play around with these settings. The little preview window will show a black and white sample. The 2 fill settings determine how crisp the outline will be. You basically want your subject to appear all white and the green area to appear solid black; avoid shades of gray.

The choke allows you to bite into the cutout to remove jagged edges. Once you think you have the settings right, click the Apply button. It will take several minutes or longer to render your shot. Once it's done, you can watch your clip. You may need to go back and change the settings a few times to get the best results.

3b. In Video Edit Magic, place your background video in the Video 1 timeline. Place your foreground video (with greenscreen) in the Video 2 timeline. Click the Video Transitions tab in the Collection window and drag Chroma Key Color to the Transition timeline. In the window that pops up, you can click your green background to sample it, and drag the Similarity slider to adjust the tolerance.

You should see the green vanishing to reveal your new background. Once you have it the way you want it, stretch the transition to the desired time span, then render and save.

Other software will take you through steps very similar to those described above. To learn more greenscreen theory, try philipwilliams.com/green screen.aspx. For an instructional video dramatizing the greenscreen process, check out makezine.com/go/green.

There's really no need for dull video backgrounds when you can key your own!

Bill Barminski is an artist, videographer, and lecturer in the Film Department at UCLA.

Quick Bits

Tips and tools for digital diversions.
By Charles Platt and Mark Frauenfelder

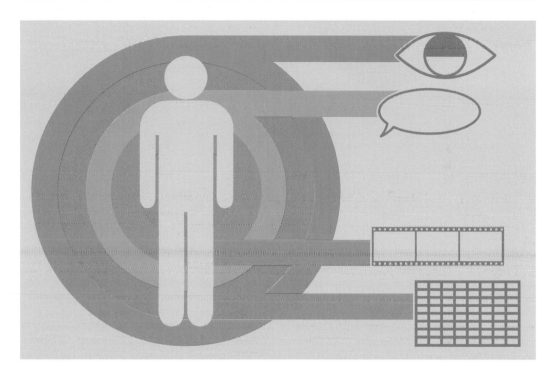

Illustration by Tim Lillis

OFF-THE-SHELF INFRARED

If you want to shoot mood photos like those in Richard Kadrey's feature on page 50, but you lack the time or patience to seek out vintage cameras with unimpaired infrared capability, an exciting new option is the Fuji Finepix IS-1.

Unlike any camera of comparable quality, this 9-megapixel digital SLR is being marketed with its IR sensitivity fully active. Simply screw in a deep-IR filter such as the Tiffen 87, and you exclude almost all visible light, revealing the infrared realm. For regular color, you substitute an IR-cut filter to block the infrared and display the world as you normally see it. Thus, this is two cameras in one.

The IS-1 is so sensitive to IR that you can use almost-normal shutter speeds, enabling photos of people and animals as well as landscapes. Since it also takes good conventional pictures, this is an interesting choice if you're thinking of buying a new camera anyway. Expect to pay slightly more than for a comparable digital SLR: around $800, plus extra for filters.

VOX POPULI

Google's Blogger software has become ubiquitous, but if you want something that looks a little different and offers multimedia functionality, check out Vox. It can help you to incorporate audio, video, and photos into blog entries.

THE FUJI IS-1 INFRARED-SENSITIVE SLR has a 10.7x lens (35mm equivalent range of 28mm to 300mm) with manual zoom. The preview LCD screen is 2" diagonal with 235,000 pixels; the electronic viewfinder also allows infrared previews.
VOX.COM is an easy way to create and maintain a multimedia-friendly blog to share your photos and videos.

After signing up at vox.com, click Design to choose a layout and a theme. Create a profile for yourself, upload your photo, and get ready for some interestingly advanced features. You can illustrate your blog text with photos from your computer, your Flickr account, your Photobucket account, or iStockphoto.

Similarly, you can merge music or spoken word originating from your computer or from Amazon. If it's your own audio, Vox limits it to 25MB but embeds it in a player that enables anyone to hear it by clicking on it. If you choose music from Amazon, Vox simply links to that source.

You can add video from your computer, Amazon, YouTube, or iFilm. Again, no video is copied from Amazon; Vox just creates a link. If it's your own video, you're limited to 50MB.

Lastly, you can package sounds, images, and videos as a "collection" that visitors open from inside your blog entry. The collection is composed of items you've uploaded to your Vox library, encapsulating the sights and sounds of a vacation or special event.

Digital devices enable us to record the world around us. Online services such as Vox offer unprecedented power and ease for sharing our experiences with anyone in the world.

SMALLEST, SIMPLEST, TOUGHEST CAMCORDER

While most camcorders still save their images onto tapes and discs that are controlled by moving parts, Sanyo has taken the radical yet obvious step of substituting flash memory for storage. Eschewing motors and bulky media, the Xacti VPC-CG65 is almost as small as a cellphone and offers solid-state reliability while making negligible compromises in image quality or recording time.

The Xacti's enhanced version of MPEG-4 looks acceptable on a full-size TV and is more than adequate for the web. Uploading video to your computer couldn't be easier: just use the supplied USB cable and click an icon to initiate the transfer.

The elimination of tapes and discs frees you from writing titles on little stickers, trying to find a safe place to store the media, and trying to find them when you want them. Your videos are saved on your hard drive along with your digital photos, and you share them with friends by emailing them as attached files.

The camera's small size, and its imperviousness to everyday abuse, can change your preconceptions about video. It isn't just something you do on special occasions anymore. You can throw your camcorder

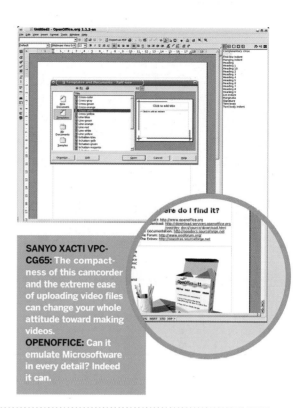

SANYO XACTI VPC-CG65: The compactness of this camcorder and the extreme ease of uploading video files can change your whole attitude toward making videos.
OPENOFFICE: Can it emulate Microsoftware in every detail? Indeed it can.

into a bag or backpack and use it with the same spontaneity as a digital camera.

Despite its tiny size, the camera's thumb-operated buttons are easy to use and intuitively laid out. You can record more than an hour of highest-quality video on an 8GB SD card, and snap individual 5-megapixel pictures along the way. An HD version is available, but for regular 640×480 video the VPC-CG65 retails for well under $400.

GETTING MORE THAN YOU PAY FOR

Software to create digital arts and crafts can be hideously expensive, but it doesn't have to be. In fact, some powerful programs are completely free — and they won't crash unexpectedly or bring malware onto your hard drive.

First and most basically, if you wince at the thought of paying a couple hundred dollars for Word, Excel, or PowerPoint, then Sun Microsystems feels your pain — and is happy to alleviate it with their open source office suite, OpenOffice.

Can it really emulate Microsoftware in every little detail? Indeed it can. Are the files really, truly compatible? So far as we can tell, they are. If you've recoiled in horror from the nightmare of Word 2007, which has taken the bizarre step of eradicating

conventional menus on the deranged theory that this will somehow make everything "simpler," maybe it's time to uninstall your MSware and embrace the open source alternative. OpenOffice runs under Windows, Mac OS X, Linux, and even Solaris (it's Sun-sponsored software, after all).

The massive downloadable package includes Calc (bearing an eerie resemblance to Excel), Writer (much like Word), Base (oddly similar to Access), Impress (like PowerPoint), and more. Some entre-preneurs on eBay will try to charge you a fee for these downloads, but you'd be foolish to pay even $4.99 to such opportunists. Just go to openoffice.org.

OpenOffice can't help you if you want to use FTP to upload data to a distant server hosting your web page, but another open source utility can handle that: a quarter-million people have downloaded WinSCP version 4, even though they would be hard pressed to pronounce its name. This freeware has an easy drag-and-drop interface, and the code seems stable. The only snag is that WinSCP runs solely on Windows.

Many people feel uneasy about downloading software online, but if you choose it selectively, it can be a very valuable alternative.

The Family Photo Archive

Use simple, powerful tools to rescue your photos from stored obscurity and turn them into a DVD slide show. By Brian O'Heir

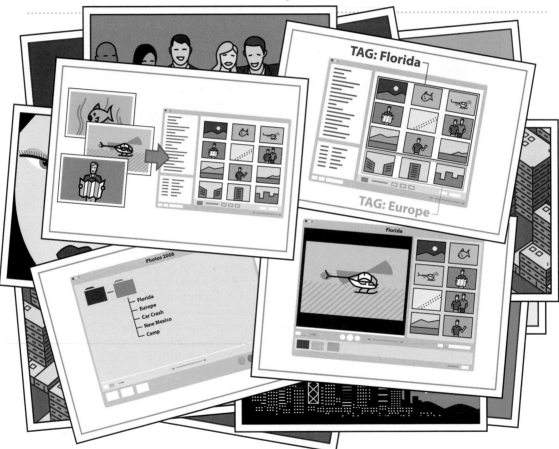

I have a wife and two children, and between the four of us, we own three digital cameras that generate literally thousands of pictures every year. These images are now piling up on my hard drive in exactly the same way that those old 4×6 photo prints of holidays, sporting events, travels, and disasters piled up in my closet. What's the point in this accumulation, if no one ever looks at any of it? Is there an easy way to transform it into a digital archive that will have enduring value? Can user-friendly software make the job tolerable? I decided to find out.

Illustrations by Nik Schulz

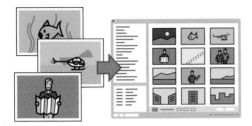

1. PREPARE SOFTWARE AND MUSIC FILES

1a. Choose a format.

I decided to turn my mess of digital photos into numerous high-definition (1080p) music-video slide shows. These would be short and fun to watch, and I could place all of them on a single DVD with a menu system allowing any to be selected and viewed by different people at different times.

1b. Get the necessary software.

I needed some really easy-to-use software, and didn't want to pay too much. As it happened, I didn't have to pay anything at all. Having recently upgraded to Windows Vista, I found that it comes bundled with some significantly improved tools to deal with digital media, most notably Windows Photo Gallery, Windows Movie Maker, and Windows DVD Maker. (Earlier versions of these programs also exist for Windows XP — and on the Mac, of course, you have tools beginning with i, such as iMovie and iPhoto.)

1c. Select background music.

I'm sure that most people have been ripping and burning audio CDs for years, but not me; all my music is in a Sony 300 CD changer. So I grabbed a dozen CDs and ripped them in Windows Media Player. This was really easy. From the Menu bar I selected Rip ⇒ Rip CD Automatically When Inserted and Eject CD After Ripping. Then I just had to insert the CD, wait for it to eject, and insert another. This was so simple I ended up ripping 287 CDs while I was doing other things on the computer.

1d. Devise a work flow.

I used Photo Gallery, Movie Maker, and DVD Maker sequentially. Photo Gallery enabled me to select and organize photos into affinity groups. Movie

PHOTO GALLERY. **Fig. A:** Photo Gallery main window showing general layout with Menu/Button bar, Navigation pane, Content pane, Info pane, and control panel. Selected menu Thumbnail ⇒ Group By ⇒ Tag, Descending. **Fig. B:** Photo Gallery main window with additional tags being added to selected photos from the group of photos with the tag "2006 DVD" (highlighted with all its subtags in the Navigation pane).

Maker promised to produce high-def animated musical slide shows from each Photo Gallery group. DVD Maker, as its name suggests, would burn them onto a disc with the menu system that I wanted.

TAG: Florida

TAG: Europe

2. ORGANIZE THE PICTURES

I started with my thousands of digital photos. Photo Gallery divides the screen into 4 sections: a central photo thumbnail viewing pane called List View, a Navigation Tree on the left, a photo Info Pane on the right, and a Media Player-type Navigation Bar control panel at the bottom (Figure A). This allows you to organize and sort photos using properties called Tags. Tags are like group names and are easily created by selecting the Add Tags button in the photo info pane. I created the 2006 DVD tag and 14 subcategory tags representing the major events of the year:

2006 DVD
    Canyon Ridge Fire
    Canyon Ridge Utilities
    Fool Cats
    House Rock
    Car Crash
    Tara Europe
    Rockport 2006
    New Mexico
    Tara's 16th Birthday
    Ian Camp
    Flagstaff Skiing
    Sedona Hiking
    Disney World
    Florida Keys Sailing

I nested all 14 subcategory tags into 2006 DVD by simply dragging them to that tag, as if I were dragging files into a folder.

To display all the photos on my hard drive in the List View pane, I clicked the button All Pictures and Videos in the Navigation Tree. I then sorted photos using the option Thumbnail ⇒ Group By ... in the menu that drops down beside the Search window.

Holding down the Ctrl or Shift key, and scrolling with the mouse wheel, I selected every picture

I wanted for the DVD, and tagged them to 2006 DVD by selecting Add Tags ⇒ 2006 DVD in the Info Pane on the right (Figure B).

Next, I clicked the 2006 DVD tag in the Navigation Tree on the left side and displayed those 1,287 images. I also applied the other 14 tags to the corresponding pictures by Ctrl-clicking pictures and selecting Add Tags followed by the appropriate tag in the pulldown menu.

This sounds complex, but it was easy to do since most pictures were already in large chronological groups. When the tagging was finished, I changed the grouping by selecting Thumbnail ⇒ Group By ⇒ Tag, Descending. This put all the various tagged groups together (they had been in different chunks because 3 of the groups were taken around the same time with different cameras), and in chronological order within each group. Now I was ready to import these groups of photos into Windows Movie Maker to design individual slide shows.

3. DESIGN A SLIDE SHOW

Most people think that good design stands out. I disagree; good design is like a transparent window revealing the subject, which in a slide show consists of the people or places in your photographs.

Good slide shows are also about emotion and imagination. They involve art more than technology. Keep it simple: good pictures and music. As the architect Ludwig Mies van der Rohe famously said, "Less is more." Simplicity is elegant. This was important to bear in mind when I used Windows Movie Maker since it has so many tempting features, which should be used with restraint.

Movie Maker divides its screen into 4 sections (Figure C). At the top center is the content pane, where audio/video/photo media Collections can be imported with the Import Media button. This content pane is also used to display Effects and Transitions. At top right is the Preview pane, like a

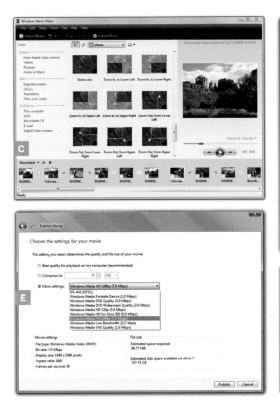

miniature Windows Media Player (or QuickTime Player).

At the top left is the Tasks pane, where you determine how you want to import, edit, and publish. At the bottom of the main window is the editing pane. This consists of the Timeline or Storyboard (they alternate, and are just different representations of the graphic media queue), the audio track, and the title overlay.

Typically, during production items (media, transitions, and effects) are dragged from the content pane to the editing pane, and the production progress is monitored in the Preview pane.

Movie Maker is designed for video but works very well with still photos for creating animated slide shows. Making each slide show involves 4 steps: selecting music, photos, transitions, and effects.

3a. Select the music.
The foundation for each slide show is a piece of music appropriate for the subject of the pictures. Popular tunes work well because they bring up remembered feelings. Music with lyrics creates a more passive experience, requiring less imagination, as lyrics solidly set the theme of the show.

Slide shows without lyrics can be more interesting with repeated viewing because the viewer is free to reinterpret each show. I used Windows Media Player running concurrently with Photo Gallery so that I could sample musical selections while I was looking at the group of photos for that particular slide show, until I found something appropriate.

Image duration is the length of time the image is seen, including transitions. It's important that the tempo of the music matches the tempo of the image duration. The default image duration is 7 seconds. It can be reset in a dialog box (Figure D) invoked by selecting from the Menu bar Tools ⇒ Options ⇒ Advanced. I decided on a default duration based on my music selection, and divided the total time of the music (in seconds) by this duration time to calculate the total number of photos I needed in the show.

Thus, picture subject determines music, music determines duration, and duration determines number of pictures.

3b. Select the photos.
One of my slide shows is of my family's sailing lessons in the Florida Keys, and the music I selected for this is "Kokomo" by the Beach Boys. I might have used

something more sophisticated but this show is mostly for my kids.

I created a New Project in the File menu and used the Import Media button to import "Kokomo" into the content pane. I dragged the "Kokomo" icon down to the Audio/Music line of the Timeline. The Timeline now reveals that "Kokomo" is 220 seconds long. I needed to set the default duration time before placing photos, so I selected Tools ⇒ Options ⇒ Advanced tab. "Kokomo" is fast enough that I left the duration at 7 seconds, producing a total photo number of 220/7 or 32.

I went back to Photo Gallery and clicked on the Florida Keys Sailing tag in the Navigation Tree, bringing up all of those photos. Holding down the Ctrl key, I individually selected 32 photos while Photo Gallery kept count. I then clicked the Make a Movie button in the Menu/Button bar, and Windows automatically added these photos to Movie Maker, inserting them into the Timeline.

3c. Choose transitions.

Movie Maker contains 2 powerful editing features: Effects and Transitions. Double-clicking on any Effect or Transition in the content pane will demo that feature in the Preview pane.

Back in the days of Kodak Carousel projectors, professional slide shows were made with multiple projectors and dissolve units. Dissolve units controlled the power to the projector bulbs, fading them on and off slowly to blend pictures from one to another. In a similar way, Movie Maker uses Transitions to blend one picture to another. The Fade Transition mimics a dissolve and is the smoothest and least disruptive way to transition from one picture to the next. I prefer it over all others.

Most of the other Transitions are major distractions. Many are very slick and impressive, and they work smoothly, but they tend to distract the viewer from the photos. This is a good thing only if your pictures are bad.

The easiest way to add a transition is to drag it from the Transitions content pane to the transition icon between 2 images on the Storyboard. To add the same Transition to many images, 1) click on the Transition and copy (Ctrl-C), 2) click the transition icon in the Storyboard, and 3) paste (Ctrl-V). Then repeat 2 and 3 as needed.

3d. Choose effects.

The real power of Movie Maker to create a slide show comes from its Effects option, particularly a group of 15 dynamic zoom and pan effects. Used in conjunction with high-resolution images (greater than 3 megapixels) these effects add flow and movement to the show. You can use them to pan across vast landscapes, zoom in on a child's face, or zoom out to reveal the whole story of a particular picture.

The easiest way to add an effect is to drag it from the Effects content pane to the image in the Storyboard or Timeline of the edit pane. To add the same effect to many images, 1) select it and copy, 2) click the star on another image, and 3) paste.

The 15 dynamic Effects should be applied so they best match the composition of each photo. Zoom in on important detail. Zoom out from detail to reveal wide-angle views. Pan to detail, not away from it. Effects should be carefully selected for each photo.

Any picture can contain multiple static or dynamic effects for a wide range of different image qualities. If a picture contains multiple effects, right-clicking the Effects star on that picture brings up the Add or Remove Effects window. Here you can change the applied Effects and their order, significantly altering the final picture. You can also copy entire collections of Effects from one picture to the next with the right-click menu items Copy and Paste.

3e. The wrap.

When my movie was complete, I clicked the Publish Movie button, which opens a dialog box that is mostly self-explanatory (Figure E). In this box you can specify different formats of your movie, for different devices. The Publish button creates the movie by saving it on your hard drive.

4. BURN THE DVD

I still needed to use DVD Maker to consolidate multiple slide show movies onto a DVD that would be compatible with any DVD player or computer. While the movies were saved in HD, I had no HD DVD burner to make an HD DVD. I resolved to make a regular-definition DVD and burn an HD version when I upgrade my equipment sometime in the future.

Windows DVD Maker is a simple, straightforward, and rather linear program. Click the Add Items button to choose photos and videos in the Add Pictures and Video to the DVD window (Figure F). Remove items by selecting them and pressing the Delete key. Rearrange files by dragging them up or down.

A pie chart indicates time used on the DVD, and a disc title window is located at the bottom, with the date as a default title; type your title here. Click the Options button to open the DVD Options window, where playback settings, aspect ratio, video format, and DVD burner speed can be set.

Click the Next button to open the Ready to Burn Disc dialog box (Figure G). At this point you can select a menu style for your DVD. A static image of each selected menu style is shown in the main window, but since most menus are dynamic, it's necessary to use the Preview feature to fully understand each style. Menu styles, and the entire DVD, can be previewed using the Preview button on the Menu/Button bar. Depending on your computer, previews can be somewhat rough, but this is no reflection of the finished DVD operation.

My finished DVD contains 14 separate musical slide shows, each lasting between 2 and 7 minutes. The DVD will play all slide shows sequentially, or individually from the menu. The slide shows are fun and simple, and popular with my family and friends. Best of all, they were quite easy to make, and now that I've climbed the learning curve, I can repeat the process easily whenever I accumulate a new batch of digital photographs. Instead of a disorganized stash of half-forgotten photos, I now have an item of genuine archival interest and value.

As for my closet full of old 4×6 color prints — they're still there. Dealing with my digital collection turned out to be much more fun.

Brian O'Heir is an architectural designer and home decor store owner living in Sedona, Ariz.

DVD MAKER. Fig. F: The Add Pictures and Video to DVD window, showing general layout, slide show queue in Content pane, Menu/Button bar, and disc title window.
Fig. G: Ready to Burn Disc window, showing Preview pane, Menu Styles pane, and Menu/Button bar.
Fig. H: Change Your Slide Show Settings window, showing music content pane and buttons, length data, picture length (duration) with auto picture length selected, and transition and effects selected.

Gnarly CAs:
Cellular Automata for Pattern Creation

Autonomous software bots can create complex, colorful digital patterns. You just have to tell them what to do. By Rudy Rucker

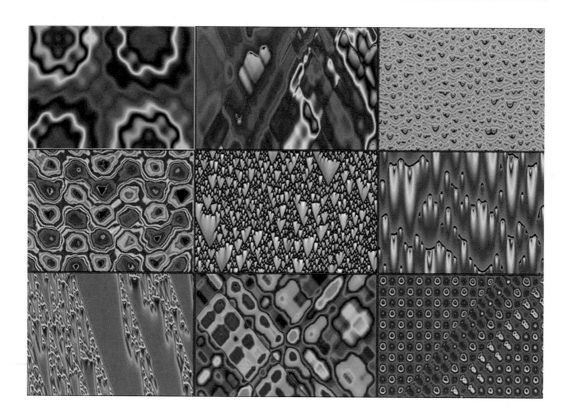

Cellular automata (CAs) are single-celled digital life forms that grow inside computer memory like bacteria, creating complex, beautiful patterns that you can copy-paste onto web pages and even onto clothing. They're easy to make, because they literally make themselves.

I find CAs fascinating in the same way that I used to love lava lamps — or blobby analog light shows in rock concerts during the late 1960s and early 1970s, where images were generated by a projector shining through mixtures of colored oil and water in a glass dish.

When I moved to Silicon Valley in 1986, I looked up the ur-hacker hero Bill Gosper. He was gloating about having given a talk using a fabulously expensive computer projector called a Light Valve, which created its images — he claimed — by beaming a carbon arc-light through three screens that were coated by ever-changing sprays of colored whale oil! Today when I speak in public, I don't need to lug around bottles of colored oil to make my light show. I can generate a shifting colored display of CAs just by hooking a computer projector to my laptop.

SNAPSHOT OF A CELLULAR AUTOMATON: A lacy, organic, 2D CA pattern of scrolls, generated by a Turing-style rule that models activator and inhibitor chemicals. The patterns emerge from a random start.

1. DOWNLOAD THE SOFTWARE

The CA program I use is Capow, which I developed with successive teams of my computer science students at San Jose State University in the 1990s. (We were funded in part by EPRI, the Electric Power Research Institute of Palo Alto.) This software is free. You can download the Windows version from rudyrucker.com/capow. This primer will tell you how to use it.

2. SAMPLE SOME PATTERNS

Capow comes with a large Help file full of suggestions. Two quick tips for getting started: open the File ⇒ Randomize dialog box from the Menu bar, leave it open to one side, and repeatedly click the Autorandomize Once button. And try opening some parameter files: *1D Lava Lamps.CAS* and *2D Life is a Gnarly Computation (pattern).CAS* bring up a lot of nice patterns that grow in one and two dimensions.

Because the program runs so rapidly, it's easy to experiment until you find just the right design for your purposes — and then it's just a matter of doing a screen capture. In Windows, you hold down Shift and press the Print Screen key (sometimes labeled Prt Scrn) to copy the screen image to the clipboard. Then open photo processing software such as Microsoft Paint or Photoshop. Create a new, blank image file and paste the clipboard image into the window.

3. EXPLORE THE POSSIBILITIES

My student Alan Borecky had the idea of making clothes from CA patterns. Alan used a CA that emulates a two-dimensional (2D) wave like you'd see on the surface of a pond. Capow's virtual 2D wave runs in a little *Asteroids*-like world that wraps around from left to right and top to bottom, so you can tile them together with no visible seams.

Borecky bought some rolls of fabric online for use in inkjet printers. He printed strips as wide as a sheet of paper, washed out the extra ink, and hung them to dry on his shower bar.

Then his wife, Donna, took over. First she made some test dresses for a Barbie doll, and then she made herself a dress. Alan was working as a game programmer at Electronic Arts at the time, and Donna wore the dress to the company Christmas party. Alan says the fellow employees they sat with were very impressed. I asked him if anyone at the

Illustrations by Rudy Rucker

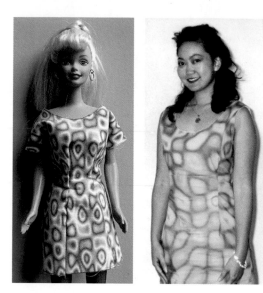

CA IN THE REAL WORLD: Barbie and Donna Borecky wearing dresses with 2D CA wave equation patterns; Donna sewed the dresses. The author's webzine *Flurb* gives each story a custom CA-generated sidebar, cultured in a 65×1200 pixel world. **FACING PAGE:** A 3D representation of a 2D CA of the Belousov-Zhabotinsky type. Cell intensities are represented as heights.

dinner suggested dressing EA characters in CA clothes. "Well, none of us had any say about things like that," said Alan. "I was sitting with all engineers."

I use CA patterns for making the page borders for my webzine Flurb (flurb.net). I run a CA program in which the display area is only 64 pixels wide, then I save a number of narrow images to use as page borders in my zine. CAs adjust themselves to whatever space you give them to compute in, so the patterns fit nicely into the narrow rectangles. My idea was to procedurally generate something like the borders that used to be on the edges of the early *Mad* comics and the late *Weirdo* comix.

THE UNDERLYING PRINCIPLES

At this point, I should explain a little more about what a cellular automaton is. The idea is to fill a region of space with identical cells, each holding a smoothly varying number, called its value. To each value, we assign a color. The tricky part is how we update the cell values. A computer program tells each cell to look at its present value relative to the values of its nearest neighbors, after which it applies a simple rule to adjust itself to its environment.

What kind of rule? One of the simplest and most interesting rules is to have the cell average its value with the values of its neighbors, add some small increment, and if the result is larger than, say, 100,

drop back down to 0. This produces a dynamic pattern that can resemble a lava lamp!

Some readers may have heard of John Conway's Game of Life, an early CA program in which the cell values are either 0 or 1. For a nice Java implementation, see math.com/students/wonders/life/life.html. In the Capow CAs, we allow the cells to have a continuous range of values, which is why they're so good at modeling nature.

I work with both one-dimensional and two-dimensional cellular automata, that is, 1D CAs and 2D CAs. In a 1D CA we're looking at something like a vibrating string, and in a 2D CA we're looking at something like a vibrating membrane.

When displaying a 1D CA, we fill up the screen by showing a kind of space-time diagram; we show successive generations of the rule, one atop the other, with the upper rows representing successively earlier generations.

When displaying a 2D CA, we're simply looking down at a little plane of virtual computers working away like mad in parallel. In Capow, you can put the cursor into "touch" mode and click on one of the images, as if throwing a rock into a virtual pond.

CAs are a good match for the physics of the natural world: both are parallel processes based on local interactions. When you set off ripples in a pool, nature is doing something very much like running

Photography by Alan Borecky (left and center) and Rudy Rucker (right)

When displaying a 2D CA, we're simply looking down at a little plane of virtual computers working away like mad in parallel.

a CA rule. Each little square of the water's surface can be thought of as having a height and a velocity variable, with each square updating its state on the basis of its height and the heights of the neighboring regions. When you pluck a string, something similar is going on, except here the elements of the computation are short segments of the string rather than little squares of a surface. In both cases, the process is obeying a differential equation known as the *wave equation*.

In order to simulate a wave in a CA, we think of each cell as holding a real number C corresponding to an intensity or, if you will, a height. Have each cell compute its new height, $NewC$, by taking the average of the neighbors' heights and adding to this a kind of velocity term equal to the difference between the cell's current value C and its previous value, $OldC$. Although it isn't obvious — at least not to me! — that this simple rule will, in fact, re-create wave motion, it works very nicely. In symbols:

(Wave Rule)$NewC = NeighborhoodAverage + C - OldC$

To make things funkier, I've found ways to make the spring-like force that drives the wave equation take on a quadratic-power or even cubic-power law, instead of the usual linear Hooke's law.

At the end of his life, computer science pioneer Alan Turing was trying to show how the patterns on animal coats and butterfly wings might emerge from CA rules based on two competing chemicals, an activator and an inhibitor. The "Turing patterns" that emerge from these rules can look like spots or filigrees.

My personal favorites are the rules that generate scrolls which biochemists call Belousov-Zhabotinsky scrolls. The dynamics of these patterns are lovely, the spirals are constantly turning, and the scrolls expand, swallow each other, and spawn off new scrolls — almost like living creatures.

If you zoom in on a 2D CA in Capow, you can use the Control ⇒ 3D View Controls ⇒ Flip button to display the pattern as an undulating three-dimensional surface. It would be great to make a surfing game based on these Belousov-Zhabotinsky scrolls, or perhaps on cubic or quadratic waves.

These days I prefer writing science fiction to writing computer code. So rather than create the ultimate surfing game, I wrote a story about it with Marc Laidlaw called "The Perfect Wave." It will be appearing in *Asimov's Science Fiction* magazine. Gnarly, dude!

Rudy Rucker is a science fiction writer, mathematician, emeritus professor in computer science, and blogger at rudyrucker.com/blog.

Remake Your Own Hollywood Movie

Dissatisfied with the director's cut? Direct it yourself!
By Richard Kadrey

Pod Race

Destroyer Droids

Midi-chlorians

Are you sure you want to delete the clip: "Annoying new character"?

no   absolutely!

Well-Delivered Dialogue

Really, it's all George Lucas' fault. By 1999, when *The Phantom Menace* was released, two generations had grown up regarding *Star Wars* with a kind of religious awe. A lot of these hardcore fans were disappointed with *The Phantom Menace*, and by the time *Revenge of the Sith* was released as the last film of the new trilogy, they felt betrayed. This wasn't the first time fans had felt ripped off by a movie series, but this time, they had the tools to do something about it.

With video editing programs illegally copied or purchased from a software store, they dropped Lucas' shabby effort onto their hard drives and went to work.

The technical part of editing wasn't difficult. Once the film was on their disks, they could even use something as basic as iMovie to slice and dice the film into individual scenes, leave some on the cutting room floor, rearrange others, and then put them back together.

BANISHING MR. BINKS

The first fan-edited movie to make a splash was known as *The Phantom Edit*, a shorter, punchier version of the original film. It exploded across the geek movie world via bootlegs and later BitTorrent — a savagely efficient global distribution system.

In *The Phantom Edit*, long stretches of Lucas' talky political scenes were eliminated. The most radical change, however, was that the almost universally loathed Jar Jar Binks suffered a digital

Illustration by Tim Lillis

"extraordinary rendition" and was virtually excised from the movie.

The radical surgery was performed by someone known at the time only as The Phantom Editor, though he was later revealed to be professional film editor Mike J. Nichols. Not only did Nichols re-edit Lucas' movie, he added commentary to his cut justifying his revisions. You can still download *The Phantom Edit* at such torrent sites as mininova.org and torrentreactor.net.

The Phantom Editor's bold move set off a wave of fan film remixes. There are now dozens of versions of the most recent Star Wars trilogy floating around the BitTorrent world. Some simply tweak the film; some alter not just the running time and chronology of the scenes, but contain technical contributions such as enhanced sound effects.

EDITS UNLEASHED

In response to these fan edits of his movies, Lucas released 250 movie clips from the Star Wars series on an official mashup area of starwars.com.

Powered by idiot-proof drag-and-drop editing tools provided by eyespot.com, fans are encouraged to edit the clips in any way they like. It's an acknowledgement that fan edits are here to stay, but it also smacks of a kind of avuncular desperation. Even a notorious control freak like Lucas must know that

trying to limit fan edits by choosing which parts of the films to make available is pretty much the definition of too little, too late.

Having whetted their appetites on Star Wars, fans have now started in on other movies and series. *Wizard People, Dear Reader* is a reworking of *Harry Potter and The Sorcerer's Stone*, the first Harry Potter movie, but this time only the soundtrack is altered. Brad Neely, a comic artist from Austin, Texas, rewrote *The Sorcerer's Stone* in his own demented and abusive style. Voldemort is renamed Val-mart. At one point, Harry refers to himself as a "beautiful animal," and his chunky cousin, Dudley Dursley, is redubbed Roast-Beefy O'Weefy. Neely's version of Potter is a profane, surreal experience, which wanders off into tangents, tall tales, and weird asides. *Wizard People, Dear Reader* doesn't follow the Potter book or movie, but is its own unique and "beautiful animal." It too is available at numerous torrent sites, such as myspleen.net.

Other fan edits available online are less invasive, such as an effort by Crappy Logo Productions in which the three *Evil Dead* movies are spliced together into one huge zombie epic.

A fan identified as Aztek463 has produced perhaps the most existential fan edit that I'm aware of, a reworking of Quentin Tarantino's *Pulp Fiction* in which the movie is dismembered and reassembled

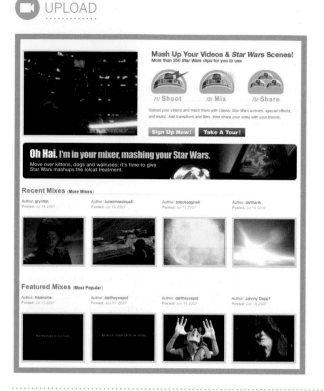

Mash Up Your Videos & *Star Wars* Scenes!
More than 250 *Star Wars* clips for you to use

/1/ Shoot   /2/ Mix   /3/ Share

Upload your videos and mash them with classic *Star Wars* scenes, special effects, and music. Add transitions and titles, then share your video with your friends.

Sign Up Now!   Take A Tour!

Oh Hai. I'm in your mixer, mashing your Star Wars.
Move over kittens, dogs and walruses; it's time to give
Star Wars mashups the lolcat treatment.

Recent Mixes (More Mixes)

Author: grymrbn   Author: lucasmaximus5   Author: bobobaggins8   Author: darthari6
Posted: Jul 19 2007   Posted: Jul 19 2007   Posted: Jul 19 2007   Posted: Jul 19 2007

Featured Mixes (Most Popular)

Author: insanoma   Author: dartheyespot   Author: dartheyespot   Author: Johnny Depp7
Posted: Jul 11 2007   Posted: Jun 57 2007   Posted: Jun 19 2007   Posted: Jun 19 2007

This wasn't the first time fans had felt ripped off by a movie series, but this time, they had the tools to do something about it.

AUTHORIZED MASHUP: George Lucas has tried to co-opt the fan editors by letting them reshuffle just a few selected clips on an "official" site (left).

into a "trilogy," separating the tangled storylines in chronological order. This utterly destroys John Travolta's feel-good strut out of the diner at the close of the original version.

Since this demented effort annihilates not just the structure, but the whole point of the original film — its disregard for chronology in favor of clever storytelling — the re-edit is both the most pointless and most brilliant fan remix ever. You can find copies at the torrent sites listed earlier.

MASHUPS AS A LITERARY TRADITION

In one sense, none of this unauthorized editing is new. A couple hundred years ago, Thomas Jefferson chopped up the Bible, taking out all the spooky God stuff in favor of what he considered the book's useful moral teachings. In the mid-1980s, John Oswald's *Plunderphonics* created mashups of other artists' songs back before the word mashup existed. Emergency Broadcast Network edited music and film clips together into evening-long dance club extravaganzas. What's changed now is that you don't need anything more sophisticated than a standard PC to create your own samizdat film epics.

It's possible that in a few years a new Tarantino or Sofia Coppola will emerge from this underground world. What better way is there to understand how

movies work than by taking them apart, seeing how the parts sync up, and then putting them back together again?

"Fan editors" of *Star Wars*, *The Matrix*, and *Harry Potter* might constitute the next generation's film school, the movie equivalent of a young guitarist hunched over in his bedroom, learning the solo from Eric Clapton's "Layla" one note at a time. And when these new filmmakers invade Hollywood to produce their own original films, you can bet that their fans will be waiting in the wings to hijack their work, remake it, and start the cycle all over again. The genie is out of the bottle, and it's about damned time.

RESOURCES

Where fan editors and their fans hang:
faneditforum.com

BitTorrent: bittorrent.com

uTorrent: utorrent.com

Azureus: azureus.sourceforge.net

Other BitTorrent sites: thepiratebay.org, mininova.org, torrentreactor.net

Boomerang
By Cy Tymony

Make a real boomerang out of cardboard and foam rubber.

You will need: Corrugated cardboard, foam rubber, tape, scissors

Have you ever wondered how birds fly or how sailboats can sail into the wind? In the last issue, I showed how to make a flying disc using the Bernoulli principle to generate lift; now let's use the same principle to create a boomerang out of ordinary stuff like cardboard.

1. Cut the cardboard to the size and shape shown. Each wing of the boomerang should be 9"×2".

2. Cut 2 foam pieces into oval shapes about 6"×2" with 1 side rising, as shown in Figure 2. The rising shape should resemble the side view of an airplane wing. Place the oval foam pieces on the leading edges of the boomerang and secure them with tape. The foam creates a curved shape on the boomerang wing, causing air to move faster across its top area than the bottom surface. This will produce lift for the boomerang.

NOTE: Look carefully at the placement of the ovals on the wings in Figure 3 before taping them.

3. Throw the boomerang like you were going to toss a baseball. Throw it straight overhead, not to the side. The boomerang should fly straight and return to the left. Experiment with different throwing angles to obtain a desired return pattern. If the boomerang does not return properly, add extra weight by using thicker cardboard or by taping coins, evenly spaced, to its surface. You can also experiment with thicker pieces of foam to improve lift.

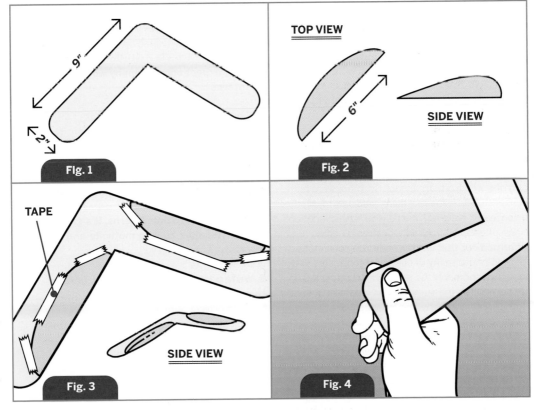

Fig. 1

TOP VIEW

SIDE VIEW

Fig. 2

TAPE

SIDE VIEW

Fig. 3

Fig. 4

Illustration by Dustin Amery Hostetler

Cy Tymony is the author of the *Sneaky Uses for Everyday Things* book series. sneakyuses.com

Making Trouble

INTERN, GET ME A CAMPARI!

WHY SUMMER INTERNSHIPS ARE MORE IMPORTANT THAN EVER BEFORE. By Saul Griffith

WITH THE LOSS OF INDUSTRIAL TRADES and craftsmanship, apprenticeships have declined steadily. They started in the Middle Ages, with young people spending about seven years living and working with master craftsmen in the hopes that one day they would become masters of their art.

The modern equivalent is the internship, devilishly satirized in *The Life Aquatic with Steve Zissou*, when Zissou asks an administrator (who happens to be a topless blonde), "Do the interns get Glocks?" The answer: "No, they all share one." It perfectly captures the two-edged sword that interns are faced with in this day and age.

This summer I had 13 interns. We rented a rambling, run-down house a short bicycle ride from the office, and the interns populated it. Two each were undergraduates from Berkeley, MIT, and Stanford, and one was from art school. Six more were high school students, most of them candidates from the InvenTeams program of the Lemelson Foundation.

Despite frustrations with California's labor laws (it turned out that interns under 18 would not be allowed to use many of our power tools, access to which was probably the reason they had signed up for servitude), the summer was a huge success, both for my company and for the students who came to work with us.

Why was I offering the internships? It's what I would have loved to do when I was in late high school. Interning in a high-tech company working on cutting-edge technology before going to college was not an option available to me, and probably not to most people. At the onset of the summer, I sat the students down to set expectations: "At worst you

won't get in the way ... At best you will make useful contributions!" It definitely turned out for the best.

I can think of only positive reasons why this should be the norm, and not the exception. I write here to encourage those readers of MAKE who have the power to offer internships to do so. More of them, lots of them. It's easier than you think. It's more rewarding than you can imagine.

SAUL'S STEP-BY-STEP GUIDE TO INTERNSHIPS:
1. Don't expect core company work to get done by interns.

2. Expect a pleasant surprise.

3. Don't underestimate the interns. They are likely smarter than you. Treat them like intelligent adults and give them responsibility over their own work.

4. If you have multiple interns, it's great to have one a little older to help motivate and manage the rest. (Thanks, Jesse.)

5. Write a long list of things to do before they arrive. I found it useful to divide the list in termsof project duration:
a. Projects of less than one day. This can be a long list. Think of all the things that don't get done around your office: tool organization, light installation, furniture building, internet research projects, etc.
b. One-day to one-week tasks. Things like helping engineers to assemble something (code or hardware), or self-contained peripheral projects.

THE FUTURE: Still life with interns, Mt. Diablo. (Or, make sure your interns have time to go sightseeing.)

Photograph by Guy Davidson

them, and demonstrate a lack of ego. Let them have pride and praise for the end result.

Our "wax on, wax off" project was a roof deck for our office, which is housed in a decommissioned air traffic control tower — not a core business need, but a wonderful thing to have. We helped the interns use CAD to design it, and you could see that after a few weeks of backbreaking, sunburnt work, they were delighted and proud that they'd engineered something from start to finish. It was beautiful and much better than I could have imagined.

9. Involve them in the brainstorming, and in the imaginative and creative parts of what you do. They don't have the biases you have. Their ideas are fresh, perhaps uninformed, but absolutely interesting and worthwhile.

10. Ask them to write a story about the things they enjoyed doing while working with you. It will make them feel good. It will make you feel great. Here's one from our intern Vicki Thomas:

"Another of the most significant things I learned this summer was how to learn from a source other than direct teaching. I spent the first several weeks feeling confused, overwhelmed, and in the way. However, I soon realized that I was learning a ton even by just observing and listening to conversations. Basically, feeling stupid is actually a great way to get smarter. I also learned how to teach myself about, and more efficiently approach, problems that I didn't know how to solve.

The last and most significant aspect of my experience that I want to include is the exposure to a real-world environment where extremely intelligent people are working, having fun, and making a difference simultaneously. I truly believe that it's important to show kids that such an environment can exist, and to inspire them to pursue ideas that create one. And let me tell you, I feel very inspired."

Thanks to all of the following interns for reminding me why I do engineering, for working hard, and hopefully for carrying this experience into their future work and giving apprenticeships back to a world that needs inspired young people more than ever before: Alex, Erich, Jesse, Josh, Garrett, Guy, Monica, Naomi, Rob, Sam, Skyler, Star, and Vicki.

c. Summer-long projects. These are non-core business projects, things that would be nice to have, or that you'd like to explore but can't justify spending your own time on.
With a list pre-generated, your ADD interns won't be able to ask the dreaded question, "What should I be doing now?"

6. Give them purchasing tasks. Let's face it, engineering is a lot about researching and purchasing components. It's time-consuming, so get them started young on how to do it well.

7. Invite them to your social events. Dinner parties you host, business networking events, public talks that are interesting, and sports activities. Our Interns vs. Employees softball game ended with many bloody knees, a drawn scoreboard, and huge smiles all around.

8. "Wax on, wax off" policy. Interns are learning what work is all about, after all. It isn't a problem to give them menial tasks, but let them know that every apprentice has had to do it at some time in their lives, including you. If you give them repetitive tasks that are arduous, do some of the work with

Saul Griffith is a co-author of *Howtoons* and was recently named a MacArthur Fellow.

Maker

Mathemagician

At 92, philosopher Martin Gardner is still exploring the perplexities of math, science, mystery, and magic.

By Donald E. Simanek

Martin Gardner's name is well known to mathematicians and scientists, as well as readers who enjoy puzzles and games. From 1956 to 1981 he wrote the "Mathematical Games" column in *Scientific American*. He has written more than 70 books, including short stories, novels, critiques of pseudoscience, and books explaining esoteric science for the layman. Approaching his 93rd birthday, he is still busy writing.

Photography by Adam Fish

Martin has written books on
magic, and contributed new
ideas to magician's magazines.
He enjoys performing tricks for
visitors, especially close-up magic
and physics demonstrations.

Maker

Milestones

Donald Simanek: You have had a long career in writing, and have won much acclaim for your work. How did you get started in writing?

Martin Gardner: My first job was as a reporter for the *Tulsa Tribune*, shortly after I got out of college. That was good training because you had to meet deadlines. I had the great title of "assistant oil editor" of the *Tulsa Tribune*. I was there for a year or so, then I went back to the University of Chicago and eventually got a job in the press relations office, writing science releases sent to local newspapers.

The first time I got any money for writing was after I got mustered out of the Navy and I went back to the University of Chicago. I sold a short story to *Esquire* magazine. It was a humorous story titled "The Horse on the Escalator," about a man who collected jokes about horses. He thought the jokes were hilarious. His wife didn't think any of them were funny, but she pretended she thought they were funny and laughed every time he told a horse joke. The title came from a joke that was going around at the time about a man who entered the Marshall Field's department store with a horse. At that time elevators all had elevator operators. The operator said, "You can't take the horse on the elevator." The man replied, "But lady, he gets sick on the escalator."

This is in a collection of my short stories, most of which are from *Esquire*, called "The No-Sided Professor." The title story was one of the earliest science fiction stories based on topology. Then for about a year I lived on sales of fiction to *Esquire* magazine — about 12 stories. When *Esquire* got a new editor and moved from Chicago to New York City, I lost my market. The new editor didn't think my stories were funny.

DS: Were you interested in science at an early age?

MG: Yes. Partly because my father was a professional geologist, with a Ph.D. in geology, who wrote many technical papers, mostly about limestone caverns. From my father I got a big dose of geology. He was also interested in astronomy. I learned from him the order of the planets from the sun.

When I was in grade school I even constructed a model of the solar system, with pictures of the planets pasted on cardboard, with a crude drawing of their orbits. So, my first interest in science was mainly through my dad's influence.

DS: Magazines such as *Science and Invention* were popular in the 1920s. Were you a fan of these?

MG: I was a big fan. *Science and Invention* was the delight of my youth, partly because of the type of articles they ran and partly because they had a science fiction story in every issue. They had a series called "Doctor Hackensaw's Secrets." Each one was a science fiction story, and editor Hugo Gernsback published 40 of them.

When he started *Amazing Stories* magazine I was a charter subscriber. I've often regretted that I didn't save the first 12 issues. I gave them all to my high school physics teacher. He was interested in reading them, and also made them available to his class. He had a big influence on me. I made very poor high school grades in history and English lit, but I got good grades in physics and math. My father bought me a copy of Sam Loyd's *Cyclopedia of 5,000 Puzzles, Tricks, and Conundrums*, and I was hooked on recreational math.

Magic

DS: You have written a huge book, *The Encyclopedia of Close-Up Magic*, and also articles for magician's magazines. When did you first become interested in magic?

MG: Again I have to go back to my father. He was not a magician at all, but he taught me a few magic tricks when I was very young, that he did very skillfully. One involved a table knife on which he put little bits of paper on the two sides. One at a time, you pretend to remove the bits of paper until the knife is empty on both sides, then you wave it and the bits of paper come back again. It uses what magicians call the "paddle move." It was the first magic trick I ever learned.

There was a trick where you put a match on a handkerchief and break the match, then open the

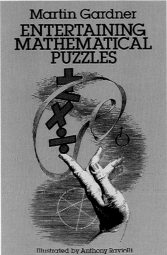

LEFT: Cover of August 1924 issue of *Science and Invention* showing imaginative Martian mining rocks. These magazines often predicted future scientific developments, but got them wrong more often than right. RIGHT: Cover of *Entertaining Mathematical Puzzles*, 1961. This picture by Anthony Ravielli captures the spirit of Martin's long interest in mathematical puzzles and games, showing the master mathemagician juggling the symbols of algebra, geometry, and topology.

handkerchief and the match is unbroken. That fooled me completely when he did it. There was another match hidden in the hem of the handkerchief. That's the one you really broke.

DS: Were you ever a performing magician?

MG: No. The only time I was a performer was when I was in college. I worked Christmas seasons from Thanksgiving to Christmas Eve at Marshall Field's in the toy department, demonstrating Mysto Magic sets. They sold a series of sets with different degrees of complexity, and there was one in particular that had very nice equipment in it. I worked out a routine and did magic behind the counter until a crowd collected. I went through a series of tricks using equipment in the magic set. I did that for about three Christmas seasons. That's the only time I ever got paid for doing magic.

DS: During your time in Chicago, did you meet any of the legendary Chicago magicians?

MG: Oh, I did. Magic was my principal hobby and I used to spend a lot of time socializing with the magic crowd. In the evenings, six to 12 magicians gathered at a Chinese restaurant called the Nankin, or another restaurant on Saturdays. Most of them worked in nightclubs. I used to take the elevated into town and join the crowd at the Nankin restaurant. Also a number of magicians who

were playing a Chicago date would appear at the restaurant as guests. So I got to meet all the local magicians and a lot of out-of-town magicians also.

Magazines

DS: You wrote the "Mathematical Games" column in *Scientific American* for 25 years, and that material found its way into 15 books. How did that job come about? Did you realize what you were getting into?

MG: No, I had no idea. I was living in New York at the time. After *Esquire* moved to New York, I realized that New York was the place for a freelance writer, so I pulled up stakes and moved to New York. I had great difficulty earning a living.

Then *Humpty Dumpty's Magazine* came along. A fellow named Harold Schwartz was in charge of their children's books. He happened to be a personal friend, and he hired me to do activity features for *Humpty Dumpty's*, which was just getting started. For eight years I was a contributing editor. In every issue I did a short story about Humpty Dumpty giving moral advice to Humpty Dumpty Junior. They were eggs of course. The magazine came out ten times a year. So I did 80 stories, which never found a book publisher.

I also did the activity features, where you do something that damages the page. You cut it, you tear it out, you fold it, you hold it up to the

Maker

"I belong to a school of thought called the Mysterians. It's a name applied to about a half-dozen philosophers. We are convinced that the human mind, consciousness, and free will are so profound and difficult to explain that no one has the slightest idea how the brain does it."

light. You make a slot and slide a strip up and down through it. These were unsigned. That was great fun. I stopped doing *Humpty Dumpty's* only when I started the *Scientific American* column.

DS: You introduced John Horton Conway's Game of Life to the world in your *Scientific American* column. This was a computer simulation game that showed the consequences of very simple rules governing the behavior of counters on a very large or even infinite grid.

MG: That was one of my most successful columns.

DS: That computer simulation shows that from just a few simple rules, complex, persistent structures with lawful and orderly behavior can and often do arise. It demonstrates that order and lawfulness can arise without purposeful design.

MG: That's one of the great lessons you get from the Game of Life.

DS: What advice can you give to aspiring writers? You once quoted Bertrand Russell's advice.

MG: He said that whenever he thinks of a simpler word that can be substituted for a complicated word, he always adopts the simpler word. I try to follow that advice.

Mathematics

DS: In 1998 you revised and modernized that classic and somewhat subversive book *Calculus Made Easy* by Silvanus P. Thompson. There's much hand-wringing amongst mathematics teachers about whether and how to reform the teaching of calculus, since many students take the course but don't really learn it well enough to use it effectively. Can you give them any advice on this matter?

MG: My only advice is to read Silvanus Thompson. I still think that's the best book to give a high school student to introduce him to calculus. It's written with a good deal of humor, and approaches the subject in a very simplified way. I think a student can learn calculus better that way than from a big, huge textbook.

DS: Probably your second-most acclaimed book is *Fads and Fallacies in the Name of Science*, which has been in print ever since 1952. It has earned you the title of "debunker" of pseudo-science. One of your books even has "debunking" in its title. Some skeptics don't like that term. They think it suggests a biased attitude and a mind that is not sufficiently open. Are you comfortable with it?

MG: Yes, I don't mind the term. I not only think it's a good term to use, but I think all professional scientists should do a certain amount of debunking in their field. A lot of them are so busy that they don't want to bother. One of the few exceptions was Carl Sagan, who wrote a couple of books you can call debunking books.

DS: You are associated with groups that have "skeptic" in their name. Can you clarify what "skeptic" means to you, and what you would define as "healthy skepticism"?

MG: As far as scientific matters are concerned, the main reason for being a skeptic is that you should not believe anything unless there's sufficient evidence.

DS: Richard Feynman said that scientific method consists of procedures we have learned to help us avoid drawing wrong conclusions.

MG: That's a good definition.

Malarkey

DS: You've written essays about perpetual motion. The idea of making a machine that puts out more energy than it takes in has been persistent since the 11th century, and shows no sign of abating.

Today some folks are pinning their hopes on tapping the "zero point energy" and "dark energy" that speculative theoretical physics are talking about. They say that the laws of physics, particularly the laws of thermodynamics, stifle creativity, and delay the time when we will produce unlimited energy for next to nothing. How would you respond to them?

MG: Well, the question of whether you can tap zero point energy is a very technical question in quantum mechanics. I'm not enough of an expert to know exactly the reasons for not believing it. So I have to base my opinion on that of the experts. Most experts in quantum mechanics believe that it's impossible to get any usable energy that way.

The leading exponent of zero point energy is Harold Puthoff, who is one of the two scientists who verified the psychic powers of Uri Geller. I have little respect for Puthoff. He's comparable to a person searching for a perpetual motion machine made of whirling wheels.

DS: Did you get a lot of correspondence from the targets of your book?

MG: I got quite a bit of correspondence.

DS: You once told me of a clever strategy you used with two folks who asked you for your opinion on their perpetual motion machine ideas.

MG: I mainly used that on angle trisectors. If I got a letter from an angle trisector I would reply, "I'm not competent to judge your construction, but you should write to so-and-so, he's an expert on it." I'd give him the name and address of another angle trisector.

DS: I'll bet you never heard from either of them again.

MG: That's true.

DS: You have also been critical about some mainstream science, such as string theory. How does a nonscientist make judgments when reading about such things? How can one draw the line between science and pseudoscience? Are there any useful clues or characteristics that one can identify?

MG: It's technically called the "demarcation problem" — the problem of distinguishing good science from bad science. There obviously isn't any sharp line. It's very difficult to decide sometimes whether a scientist is just a maverick scientist who may hit on something new or whether he's a crank. String theory is a case in point. I really have no business criticizing string theory because I don't understand it too well.

DS: Our mutual friend Bob Schadewald said that pseudosciences were entertaining, and mostly harmless, except for one. The one he thought "dangerous" was creationism. Are any others dangerous?

Maker

MG: An area of pseudoscience that's really harmful is pseudo medical science. A person can get hooked on something like, say, homeopathy. Clearly there's absolutely no scientific basis for a homeopathic drug being of any value at all because it's diluted to the point where there isn't any of the drug left. A basic dictum of homeopathy is that the more dilute the drug, the more potent it is. They dilute the drug until there are only a few molecules of it left, or not even that.

A homeopathic drug has absolutely no therapeutic effect, but if a person believes the drug is effective it can have a strong psychological effect. A person who is hooked on homeopathy can start relying on the drugs and not go to a reputable doctor, when he might have a serious illness. People could die.

DS: A while back, Jacques Benveniste [1935–2004] rationalized homeopathy by claiming that water had a memory of anything that was once dissolved in it. If that were true, then any tap water we drink would have memory of everything that was ever in it, poisons, medicines, and everything else taken out at the treatment plant. A glass of tap water would be as beneficial as a homeopathic medicine.

MG: When James Randi gives a lecture, he produces a bottle of a homeopathic remedy made from a chemical which, when taken in quantity, is poisonous. He drinks the entire bottle in front of the audience.

Mysterians

DS: Those 1920s *Science and Invention* magazines often speculated on the future of science and technology. Very often they got it wrong, failing to anticipate important innovations. You are now 92 years of age, and have seen many new things emerge from science. What scientific breakthroughs were most unexpected?

MG: Television was certainly one of them. I remember as a boy when the first television began to be available I was really awed by it. And of course, the computer revolution was another.

DS: Do you care to speculate about the next big breakthroughs in science?

MG: I suppose the biggest breakthrough in computer science would be quantum computers. They'd be so much faster they would open up all kinds of possibilities.

DS: Will artificial intelligence ever duplicate or surpass the human mind?

MG: I'm convinced it will not. I belong to a school of thought called the Mysterians. It's a name applied to about a half-dozen philosophers. We are convinced that the human mind, consciousness, and free will are so profound and difficult to explain that no one has the slightest idea how the brain does it.

DS: I think it was Von Neumann who said that if we ever make computers that can think, with the power of the human brain or better, we won't know how they do it either.

MG: Well, that's true. It may be possible that computers will imitate human intelligence in some far distant future. I don't think those computers will be made of wires and switches.

DS: What if we allow them to learn by themselves and make mistakes? Don't humans learn by making mistakes?

MG: We do. But some kind of threshold is crossed when a computer becomes aware of its own existence. The name "consciousness" applies to that. That is a major threshold that I don't think any computer that we know how to build will cross. I can imagine in a far future if a computer were made of organic material it might be able to imitate a human brain, but as long as it's made of electrical currents and switches I don't think it will cross the threshold.

Donald Simanek is emeritus professor of physics at Lock Haven University of Pennsylvania. Visit his pages of science, pseudoscience, and humor: www.lhup.edu/~dsimanek.

Make: Projects

Need to blow off steam? Toot your own horn by building our snazzy shop whistle out of PVC and compressed air. Then create sweet sol music with a sun-powered xylophone, better than wind chimes any (sunny) day. Next, exercise some remote control over your environmental impact with our kinetic remote, and make Faraday proud!

Super Tritone Shop Whistle
88

Solar Xylophone
98

Kinetic Remote Control
108

Photograph by Sam Murphy

SUPER TRITONE SHOP WHISTLE

By William Gurstelle

THIS JAZZY, COMPRESSED-AIR-POWERED WHISTLE SOUNDS A MIGHTY BLAST.

Decades ago, whistles were used in factories, on railroads, and aboard ships. At noon, whistles of every pitch could be heard informing workers that lunchtime had arrived. Railroad engineers used whistle codes for communication both within the train and with other trains.

According to the National Library of Wales, Adrian Stephens of South Wales invented the first steam whistle in 1833. Stephens was employed at the Dowlais Iron Works, where he devised a steam whistle as a warning device for boilers. The steam whistle idea took hold, and by 1860 whistles were installed in factories all across Europe and the industrial northern states of the United States. The whistles were practical and economical, and soon the steam whistle went mobile, blaring out warnings at railroad crossings.

Set up: p.92   Make it: p.93   Use it: p.97

William Gurstelle is a MAKE contributing editor and the author of several maker-friendly books including *Backyard Ballistics*, *Whoosh*, *Boom*, *Splat*, and *The Art of the Catapult*.

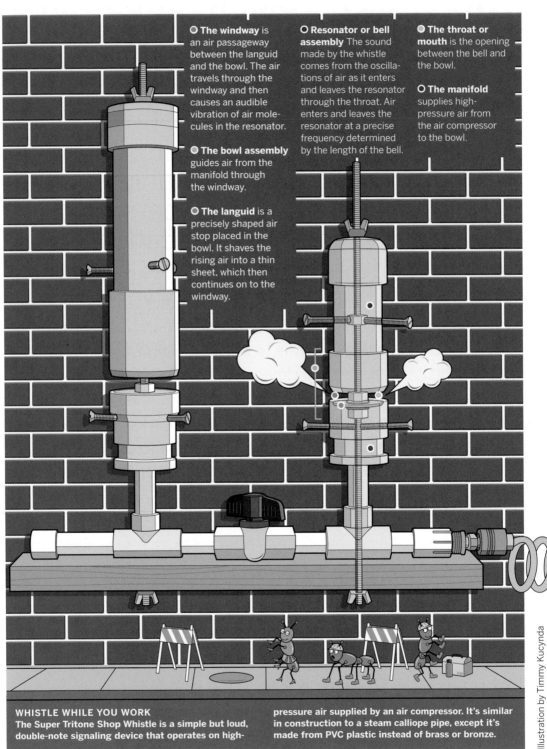

The windway is an air passageway between the languid and the bowl. The air travels through the windway and then causes an audible vibration of air molecules in the resonator.

The bowl assembly guides air from the manifold through the windway.

The languid is a precisely shaped air stop placed in the bowl. It shaves the rising air into a thin sheet, which then continues on to the windway.

Resonator or bell assembly The sound made by the whistle comes from the oscillations of air as it enters and leaves the resonator through the throat. Air enters and leaves the resonator at a precise frequency determined by the length of the bell.

The throat or mouth is the opening between the bell and the bowl.

The manifold supplies high-pressure air from the air compressor to the bowl.

WHISTLE WHILE YOU WORK
The Super Tritone Shop Whistle is a simple but loud, double-note signaling device that operates on high-pressure air supplied by an air compressor. It's similar in construction to a steam calliope pipe, except it's made from PVC plastic instead of brass or bronze.

Illustration by Timmy Kucynda

TOOT SWEET

Utilitarian shop whistles and elegant organ pipes have a lot in common.

Basically, a whistle works by splitting a column of fast-moving air with a narrow blade. Doing so creates turbulence that in turn causes the moving air column to vibrate. A resonant chamber or "bell" placed above the blade allows the whistle to be tuned to a particular frequency or pitch, while simultaneously increasing the wave amplitude (sound volume).

The layout and tolerances of a whistle are exacting. The difference of only a few thousandths of an inch in any of the components (see facing page) can mean the difference between a pure, loud whistle tone and a mushy hiss of chaotic air.

I wanted to outfit my new workshop with a loud, well-tuned compressed-air whistle. When noon rolls around, there's nothing more satisfying than pulling a cord to sound a long, sustained honk before heading in for lunch. My first few attempts at whistle making were mediocre at best, resulting in devices that did little more than hiss and chuff, emptying my air compressor's tank faster than three Green Bay Packers fans sharing a can of lager.

Frustrated, I turned to Timothy Patterson of Patterson Organworks (pipeorgan.us) in St. Paul, Minn. He's a master organ builder and a man with an unsurpassed understanding of vibrating columns of air. Patterson has worked on organs around the world, from a 400-year-old organ in Normandy to the latest computer-controlled models in the United States.

In Patterson's modest shop stands a nearly completed 15-rank, 33-stop pipe organ (shown here). The organ pipes, up to 16 feet long, share the same operating principles as my contemplated air whistle. Clearly, if anybody could get my whistle to sound properly, it is he.

Patterson and I sketched out several ideas, finally developing a design for a robust air whistle capable of producing a rock-concert-like 118 decibels at a distance of 6 feet, using a standard home air compressor.

Even better, the lengths and diameters of the two

resonators specified produce a strident, attention-demanding sound: the wonderfully dissonant interval known as the tritone. (For non-music majors, the tritone is the musical interval that spans three whole tones. Musicians would recognize the sound as an augmented fourth. Commonly employed to produce musical discordance, the tritone is well suited for a signaling whistle.)

In medieval times the tritone was called *diabolus in musica*, or more simply "The Devil's Interval." The sound of the tritone was so dissonant that it was banned from church music. In fact, some church officials of the Middle Ages believed that simply hearing the interval would stir up sexual desires in the listener. Thus, churchmen said the tritone was the sound of Satan.

The Super Tritone Shop Whistle does not seem particularly diabolical until the time comes to tune it properly. But once completed and tuned, the Super Tritone sounds with a mighty blast, letting the neighborhood know it's lunchtime, quitting time, or simply time to blow the whistle.

🎵 To hear the Super Tritone Shop Whistle, visit makezine.com/12/whistle.

SET UP.

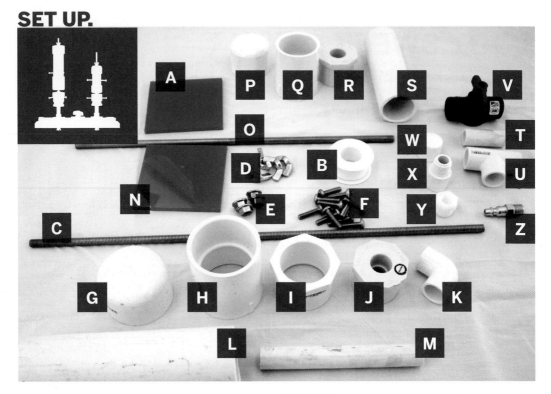

MATERIALS

FOR A WHISTLE TUNED TO F#:

[A] ¼"-thick plastic blank, 4"×4" or larger

[B] Teflon tape

[C] 20"-long, ³/8"-diameter threaded steel rod

[D] ³/8" wing nuts (2)

[E] ³/8" flange nuts (2)

[F] ¼"×1½" machine screws or thumb screws (6)

PVC PARTS:
[G] 2"-diameter smooth (non-threaded) cap fitting

[H] 2"-diameter smooth-to-smooth pipe coupling

[I] 2" to 1½" smooth-to-smooth reducing coupling

[J] 1½" to ½" smooth-to-smooth reducing coupling

[K] ½"-diameter elbow

[L] 2"-diameter PVC pipe, approximately 8" long

[M] ½"-diameter PVC pipe, approximately 6" long

FOR A WHISTLE TUNED TO C:
[N] ¼"-thick plastic blank, 4"×4" or larger

[B] Teflon tape

[O] 16"-long, ³/8" diameter threaded steel rod

[D] ³/8" wing nuts (2)

[E] ³/8" flange nuts (2) Regular nuts may be substituted if flange nuts are not available.

[F] ¼"×1½" machine screws or thumb screws (6)

PVC PARTS:
[P] 1½"-diameter smooth (non-threaded) cap fitting

[Q] 1½"-diameter smooth-to-smooth pipe coupling

[R] 1½" to ½" smooth-to-smooth reducing coupling

[S] 1½"-diameter PVC pipe, approximately 6" long

FOR MANIFOLD:
[T] ½"-diameter pipes, each 2" long (3)

[U] ½"-diameter PVC tee fittings (2)

[V] ½"-diameter ball valve

[W] ½" end cap

[X] ½" smooth-to-threaded PVC adapter

[Y] ½" to ¼" or ½" to ³/8" threaded pipe adapter

[Z] ¼" or ³/8" male industrial-style pneumatic air plug

TOOLS

[NOT SHOWN]

Hacksaw for cutting threaded rod

Plastic pipe tool for shaping PVC edges. A lathe or mill works best, but sanding devices will suffice.

Drill or drill press with ⁷/32" bit and ³/8" bit

2"-diameter hole saw for cutting the languid for the 1½" pipe

2½"-diameter hole saw for cutting the languid for the 2" pipe

High-pressure air source and nozzle

BUILD YOUR SUPER TRITONE SHOP WHISTLE

START »»   **Time: A Weekend   Complexity: Medium**

1. MAKE THE BOWLS AND LANGUIDS

Begin by fabricating the bowl and languid. Making and positioning the languid is the most exacting part of the whistle fabrication process.

1a. To make the languid for the large whistle, begin by cutting out a 2½" disk of plastic from the blank, using a hand-held power drill or, preferably, a drill press.

1b. To make the bowl, remove any bevel or angle from the 2" side of the 2" to 1½" smooth-to-smooth reducing coupling using a lathe, mill, or sanding tool. The edge must be square.

1c. Next, add the 3 bowl-positioning bolts. To do so, drill three ⁷/₃₂" holes spaced equally ¹/₃ of the way around the pipe fitting. To space the holes, wrap a piece of paper around the fitting and mark the point where the beginning and end overlap. Measure the distance between the marks and divide by 3. Note the ¹/₃ increments on the paper and then transfer the increments to the pipe fitting.

Insert 1½"-long, ¼" bolts into the holes you drilled. The ⁷/₃₂" holes are slightly narrower than the bolts, so the bolts will self-tap into the softer plastic. This allows the screws to be positioned in and out by turning them. Align the centerline of the bowl with the threaded rod, and tighten the adjustment bolts.

1d. Repeat Steps 1a–1c to make the languid and bowl for the small whistle, cutting a 2" plastic disk and using the 1½" to ½" smooth-to-smooth reducing coupling for the bowl.

2. SIZE, SHAPE, AND POSITION THE LANGUID

2a. Chuck the plastic disk to the drill using a ¼" bolt and 2 nuts. Grind the disk down to size, carefully, until there's an opening of approximately 40 thousandths of an inch between the bowl and the languid edge.

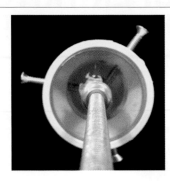 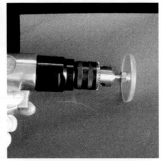

2b. To shape the languid, cut (with a lathe) or grind (with a sander) a 15° angle into the edge of the languid.

2c. Position the languid on the center post, using 2 flanged nuts. The languid should be initially positioned such that the top of the languid is even with the top of the bowl. When complete, the bowl and languid will appear as shown at right.

3. CONSTRUCT THE BELLS

3a. Begin by drilling a ⅜" hole in the center of the 2" end cap.

3b. Next, drill three ⁷/₃₂" holes spaced equally ¹/₃ of the way around the 8½"-long, 2"-diameter pipe, approximately 4" from the top of the pipe. Space the holes evenly as you did in Step 1c. Screw the ¼" bolts into the holes and position them in and out by turning them.

3c. To shape the bell's edge, cut (with a lathe) or grind (with a sander) a 30° angle into 1 edge of the 2"-diameter smooth PVC coupling as shown here.

3d. To attain the desired tritone, you'll need to cut the pipe so the length of the 2"-diameter bell assembly, from the bottom of the lip bevel to the inside top of the cap, is 8¼". Dry-assemble the bell by connecting the top cap, the bell adjustment chamber (the 2" pipe with adjusting screws), and the bell edge (the 2"-diameter smooth coupling with 45° bottom edge). Cut the pipe so that the overall length of the 2" bell assembly is 8¼" long. Disassemble after cutting the pipe to length.

3e. Now, to make the second smaller bell, repeat Steps 3a through 3d, using the 1½" pipe, end cap, and smooth-to-smooth coupling. For the tritone, cut the 1½"-diameter pipe such that the total height of the bell is 5¾" from the lip bevel to the inside top of the cap.

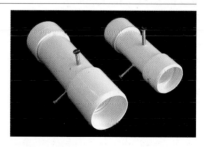

4. MAKE THE MANIFOLD AND ASSEMBLE THE WHISTLE

4a. Drill a ³/₈" hole in the bottom of each tee fitting, as shown here. Connect the manifold parts. The valve and openings at both ends allow you to tune or "voice" each whistle individually without disassembling the manifold.

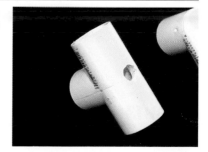

4b. Refer to the cutaway diagram and the assembled whistle drawing at makezine.com/12/whistle for a guide to final assembly. For both whistles, position a flange hex nut on the ⅜" threaded rod approximately 3" from the top. Wrap Teflon tape on the threaded rod to make the nut turn more smoothly.

4c. Place the PVC pipes with the adjustment bolts over their respective rods and turn the adjustment bolts until they contact the flange hex nut. Carefully tighten the adjustment bolts onto the flange hex nut, maintaining the pipe's vertical aspect (i.e. make the pipe centerline parallel to the threaded rod). This allows the height of the mouth or air opening to be adjusted simply by spinning the bell up or down.

4d. Place the hole in the end cap over the rod. Then place the smooth-to-smooth coupling over the bottom of the pipe. Next, place the wing nut over the center post and gently tighten. The bell edge must be positioned directly above the windway space between the languid and the bowl, and parallel to the languid. Retighten the positioning bolts if necessary to attain the correct positioning.

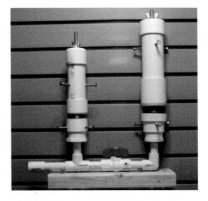

4e. Position the bowls and manifold as shown in the assembly diagram, then adjust the bells and bowls until the mouth opening is approximately ½" high.

5. VOICE THE WHISTLE

Voicing the whistle, that is, adjusting it to produce a satisfactory musical tone, takes considerable time and patience. There are a dozen variables that could be tinkered with, but we'll limit ourselves to 3 critical corrections: languid height, mouth height, and bowl/bell edge alignment. Getting it right can be difficult. So take your time and have patience. Eventually, you'll come across the correct combination to give your whistle great sound.

5a. Adjust the mouth height by spinning the bell up or down.

5b. Move the languid up or down relative to the bowl by turning the flange nuts holding the languid in place. Start with the top of the languid even with the top of the bowl and adjust up or down in small increments.

5c. For optimal bowl/bell edge alignment, keep the bell edge directly over the bowl. You can make fine adjustments by carefully tightening and loosening the bell and/or bowl adjustment bolts.

Continue to make small adjustments until you attain a loud, clear sound. Be patient, this can take a while!

FINISH ☒

NOW GO USE IT

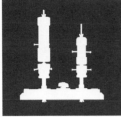

USE IT.

THAR SHE BLOWS!

AIR SUPPLY

Set compressor output to about 60psi. The pressure may be slightly increased or decreased depending on the characteristics of your whistle. Pressure that is too low may fail to produce a whistle sound. Too much pressure will result in overblowing and cause the whistle to squeak.

The simplest way to supply air to the whistle is to connect a standard industrial air line coupler to a male plug on the end of the manifold. For more control, insert a lever-operated air valve between the manifold and the air hose.

TROUBLESHOOTING

» It is absolutely critical that no air leaks out from the top where the threaded rod passes through the end cap. Even a small leak will kill your whistle's sound. Screw the wing nuts down securely, and use an O-ring or Teflon tape to maintain the integrity of the seal. Any air leak can negatively impact whistle performance. Leak-test all connections with soapy water.

» The bowl edge must be even and as close to 90° as possible.

» Make the 30° bevel on the bottom bell edge as smooth and even as possible. Depending on your whistle, the optimum bell edge angle may be a bit more or less than 30°, but a smooth, consistent cut is important.

» Make the 15° bevel on the languid as smooth and even as possible. Again, depending on your whistle, the optimum languid angle may be greater or less than 15°, but a smooth, consistent cut is important

» Different whistles behave differently at different air pressures. Experiment with different pressures until you get the sound desired.

⚠ **CAUTION: The Super Tritone Shop Whistle is an "off-label" use of the parts described here.** While I've made several of these whistles without incident, no guarantee of safety Is given or implied. The high-pressure air utilized in any pneumatic project can be dangerous to sight and hearing. Always wear safety glasses when building or operating the Super Tritone.

Four-Fingered Glove:
I broke a finger and couldn't use regular work gloves. So I removed the stitching on the two middle fingers of a glove and sewed them together to accommodate my two taped-together fingers.

The gloves looked like alien hands but worked outstandingly well. Neat side effect: they make you feel like you have more power in your hand as they focus more energy on fewer glove grip pads. —*Perry Kaye*

Find more tools, tips, and tricks at makezine.com/tnt.

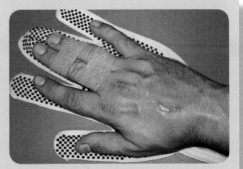

SOLAR XYLOPHONE

By Rory Nugent

MUSIC OF THE SPHERE

Solar cells gracefully link technology with the Earth's natural resources, bringing projects out of the dank, dusty workshop and giving them a sustainable home with the plants outside. This autonomous xylophone uses Solarengine circuits and pentatonic chimes to play in tune with that big nuclear power plant in the sky.

Wind chimes capture wind energy to move metal tubes that generate sound when they strike one another. They're simple, timeless, and beautiful. You can never predict the composition the chimes will play after the next gust of wind, which is what makes these inventions so compelling. The elegant overlay they add to our experiences brings us closer to nature.

I wanted to create a different kind of autonomous musical instrument that would, like wind chimes, generate tones from a natural resource. So I made this solar xylophone, which gives voice to the silent sun and takes the project (and ourselves) outside, where we belong. It uses eight simple, independent systems to strike its eight chimes in parallel. So you can lose the power cord and forget the batteries, but be sure to bring your suntan lotion.

**Set up: p.101   Make it: p.102   Use it: p.107

Rory Nugent (prize-pony.com) is a tinkerer who lives in New Jersey and loves to drink iced tea. He is currently a student at New York University's Interactive Telecommunications Program.

X-PHONE SPEX

HOW IT WORKS

A regular xylophone fits inside of a solar-powered player box that holds a mallet over each of its 8 chime tubes. Each mallet is powered by a system that includes a solar cell, a simple Solarengine circuit, and a small motor. The systems work in parallel; the brighter the sunshine on each panel, the more frequently its corresponding tube will be struck.

The sun shines upon your project.

The 8V 44mA epoxy solar cell collects light energy, converts it to DC, and trickles it into the circuit. The more light it sees, the more power it delivers.

The circuit, a Miller Solarengine, collects the energy in a 4,700µF capacitor until its voltage exceeds 5V. Then a voltage trigger opens and discharges the capacitor into the motor. A smaller 1.0µF monolithic capacitor sets the discharge's duration.

With each pulse of voltage, the pager motor pulls its mallet down with a bent paper clip. The motor has a spring on the shaft that inhibits rotation beyond 60°. This means it gives one strong momentary nudge rather than rotating continuously. This is useful for our application.

A counterweight, made of plumbing solder, makes it easy for the motor to keep the mallet head lifted off the tube between strikes.

When the mallet is pulled down, it strikes its chime tube. The 2 balance strips that run along either side of the instrument hold the mallets and serve as a fulcrum for them to swing around.

The chimes are tuned to cover 1½ octaves using a pentatonic scale. This keeps the notes sounding pleasant and not dissonant in random combinations.

Illustration by Nik Schulz

SET UP.

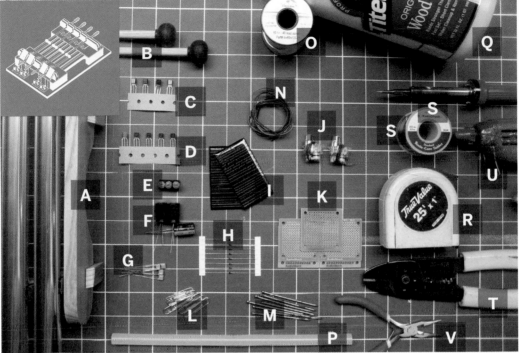

MATERIALS

[A] Xylophone I used the 8-tube Pipedream from Woodstock Chimes, $77 from chimes.com, or check Amazon and eBay. To make it less expensive, you can use smaller models with fewer tubes, even just 1 or 2, if you adjust sizes and quantities accordingly.

[B] Mallets (8) Xylophones typically come with a pair, so you'll need to obtain 6 more. Buy cheap wooden ones, which are easier to drill through. I got mine for $4 each from Woodstock.

[C] Panasonic 1381U, 4.6V voltage triggers (8) Buy 10 for a discount from solarbotics.com.

[D] 2N3904 NPN transistors (8) RadioShack part #276-2016.

[E] 10kΩ single-turn trimpot potentiometers (8) RadioShack part #271 282

[F] 4,700µF electrolytic capacitors (8)

[G] 1.0µF monolithic capacitors (8)

[H] 1N914 silicon diodes (8) Buy 10 for a discount from solarbotics.com. RadioShack part #276-1122

[I] SCC3766 8V solar cells (8)

[J] GM10 geared pager motors (8)

[K] Dual mini perf boards (4) RadioShack part #276-148, 2 boards per package

[L] Paper clips (8)

[M] 10-penny finishing nails (8)

[N] 22-gauge solid wire, about 15'

[O] Plumbing solder, about 0.125" in diameter

[P] Hot glue Epoxy will also work nicely, but takes more time to mix and apply.

[Q] Wood glue

[NOT SHOWN]
⅛" plywood, at least 12"×31"

¾" wood board, at least 16"×26" You could go less than ¾" thick, but I wouldn't suggest anything less than ¼".

1¼"×5½" wood plank, at least 14" long You could go less than 1¼" thick, but I wouldn't suggest anything less than ½".

1"×1½" wood plank, at least 31" long

Female (16) and male (8) wire connection headers (optional) These will keep your components modular and will make testing much easier.

TOOLS

[R] Measuring tape

[S] Soldering iron and solder

[T] Wire stripper/cutter

[U] Hot glue gun unless you're using epoxy

[V] Needlenose pliers

[NOT SHOWN]
Dremel tool

Drill press or drill and vise

Radial saw or other type of saw

Table saw or router and fence

Photography by Rory Nugent

MAKE IT.

BUILD YOUR
SOLAR XYLOPHONE

START ∷∷ _____ Time: **A Weekend** Complexity: **Hard**

1. ASSEMBLE THE CIRCUITS

Putting the circuits together is the most straightforward step in this project, and it will ease us into the rest.

1a. Split your RadioShack perf board into single squares, or if you're using your own perf board, cut it into 2"×2" pieces.

1b. Collect and separate the parts needed to make each circuit: voltage trigger, transistor, trimpot, diode, 4,700µF capacitor, 1.0µF capacitor, motor, solar cell, a bit of wire, 2 female connection headers (optional), and a piece of perf board.

1c. Take one group of parts and solder a circuit together, following the Solarengine schematic at makezine.com/12/xylophone. This can be a bit tricky but it's good practice if you want to get better at electronics. I put the electrolytic capacitor and the trimpot at opposite ends, and included 2 female headers for connecting to the solar cell and motor at the edges.

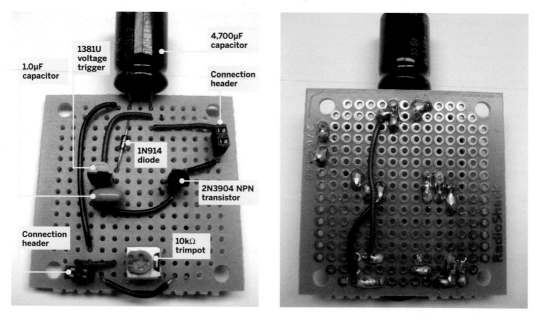

1d. After you finish one of the circuit boards, solder wire leads onto the positive and negative terminals on the solar cell. If you're using headers, solder male headers to the motor's delicate leads (we don't need these for the solar cell — the 22-gauge wire is thick enough).

1e. Test your first completed circuit board by taking it outside or putting the cell under a high-wattage incandescent bulb. You should see the motor move within 10 seconds. A compact fluorescent bulb or heavy-duty flashlight will probably also work but will take longer, and you might begin to think something is wrong with your circuit.

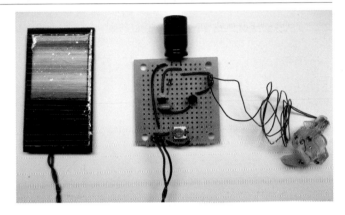

1f. If it doesn't work, adjust the trimpot using a screwdriver, by turning it all the way in one direction or the other. This controls the efficiency of the circuit, allowing more electricity to reach the capacitor, and thus, adjusting the timing of the motor. One side of the trimpot resists all the electricity while the other resists none.

1g. The circuit works correctly if the motor periodically turns about 60° and then quickly resets itself. If it works, pat yourself on the back, call up your friends for help, and build 7 more just like it (or however many your project calls for). Otherwise, check your solder connections, review the circuit, and use a multimeter on the solar cell and 4,700µF capacitor to see if electricity is flowing. Lastly, ask a friend to review the circuit.

2. BUILD THE PLATFORM AND STAND

I used a radial saw for cutting the wood and a table saw to put in some long grooves, but you can use other tools.

2a. Cut 2 pieces of ⅛" plywood to 5½"×15¼" each. These pieces will make the platforms that the mallet and motor mechanisms rest on.

2b. Cut 4 pieces of the 1¼" wood to 3¹/₁₆"×5½". These will be the legs that support the mallet platforms above the xylophone.

2c. Cut 1 piece of ¾" wood to 16"×25¾". This is the largest piece of wood and will support the entire structure.

2d. Take the 1"×1½" wood plank and cut 2 pieces 15¼" long each.

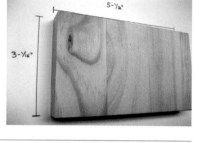

2e. This is the tricky part. Now we'll need to put a ½"-deep cut along the entire 1" edge of each 1"×1½" piece. A table saw is perfect for this. Place each piece on the table with the narrow edge down, adjust the fence so that the blade is centered on the wood, and run it all the way through the piece. Repeat for the second piece.

2f. Move the pieces back to the radial saw and cut four ⁵/₁₆" slits into each, 2¼" apart, on the same side as the long groove. These will be the notches that hold the mallets. Start at one end of the piece and measure 3³/₁₆" from the end. Mark off a section ⁵/₁₆" wide and ⅞" deep. Measure 2¼" from the marked section and repeat the marking 3 more times. Repeat this step for the second piece.

2g. Glue the 2 plywood platforms on top of the 4 leg pieces to make 2 U-shaped raised platforms. Then glue the platform tops onto the large piece of ¾" wood, running parallel and centered 2⅜" from each end. Let the glue dry if needed.

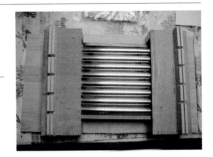

2h. Glue the 2 slotted strips on top of the platforms, running parallel and 1¼" in from the outside edges. You should now have a nice wooden structure that will house the xylophone inside, below the platforms. Each slotted wooden strip will act as a balancing point for 4 mallets.

3. INSTALL THE MALLETS

3a. Use a ¹⁄₁₆" drill bit to drill 2 holes into each mallet, one 5⅛" from the rubber ball end, and the other 4⅛" from the bare end. Slide a 10-penny nail into the hole closest to the bare end.

3b. Cut and bend your paper clips into the zigzag shape shown here using wire cutters and needlenose pliers. These wires will run through the holes closest to the mallets' striking ends and drive them from the motor.

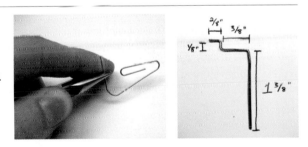

4. FINAL ASSEMBLY

This part is particularly tricky. We need to figure out where the motors should be placed and how much counterweight (plumbing solder) to spool around the mallets' ends.

4a. Slide the xylophone in place under the platforms. Use your eyes to center it.

4b. Place each mallet with its nail into its own groove on the balancing strips. The mallets should fall and rest on or near the center of each pipe. If not, adjust the placement of the xylophone underneath, or pull out and reverse the nail — the head can sometimes push a mallet off center. If that doesn't work, check the alignment of the strips and the placement of the platforms. Handcrafting an instrument like this takes some tinkering.

4c. Grab one of the mallets and wrap some plumbing solder around the wooden end. You want to wrap just enough to make the mallet balance nicely in the wooden groove and fall slowly down toward the xylophone, which requires trial and error.

4d. Place a properly counterweighted mallet into the groove, and then bring the motor from one of your circuits close to the mallet head. Fit a bent paper clip into position, connecting the motor to the mallet, with the short end through the eyehole of the motor and the long end through the unused hole in the mallet. Curve the long end of the wire up and around the mallet.

4e. Hold the motor slightly offset from the mallet, and make sure the motor is oriented so that it will pull down on the mallet.

4f. Test the circuit while holding the motor in place. This will give you an idea if you've found the right spot. If so, mark its position on the platform with a pencil. Grab your Dremel tool, attach a round wooden shaving tip, and grind out the line you marked so that the plastic tab on the bottom of the motor will fit in it. The motor will then sit flush with the platform.

4g. Repeat the installations and tests for all other mallet/motor/circuit combinations. After each motor placement mark has been ground out, do one last quick test and then hot-glue or epoxy the motor into place.

4h. Finally, drill a hole in the balancing strips next to each motor and thread the motor lead wires through. Hot-glue or epoxy the solar cells so that they rest at a 45° angle with the top touching the strip above the hole and the bottom touching the platform edge. Connect your solar cells and motors back to the circuit. You're done!

FINISH X

NOW GO USE IT »

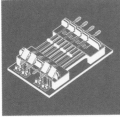

PLAY SOME SOL MUSIC

FINE-TUNING THE PAPER CLIPS

Your solar xylophone will likely need some tweaking, especially the bent paper clips. Sometimes the piece that slips through the eyehole of the motor strikes the platform. Watch closely to see if this is a problem. Make sure the long end of the paper clip is wrapped snugly around the mallet so that it grabs the mallet firmly on each strike. You may also need to shorten the tall, vertical portion of the paper clip to reduce the distance the mallet head needs to travel to sound its note.

Remember, you hand-made these motor clips; they weren't stamped out by a machine, so there is room for error. Don't be discouraged. Work at it, try different bends, and be proud you could do this all with your hands!

CONCERT ON THE GREEN

Now all the hard work is done, and you'll be relieved to hear that the solar xylophone is not difficult to operate. On a sunny day, put on your favorite sunglasses, tuck a blanket under your shoulder, and carry the xylophone out to your lawn or local park. Before you even make it to the grass, the xylophone will probably start hammering away.

If it's triggering notes too quickly for its own good, adjust the trimpot on each circuit to a more fitting speed. You may have noticed that the mallets fired at a fairly pleasant rate during your tests inside, but now that you're under the midday sun, the xylophone has become a musical monster.

You can also put it in dappled tree shade for a randomizing effect. Solar cells work best with natural, full-spectrum sunlight, much better than with artificial sources like incandescent or compact-fluorescent light bulbs.

THEMES AND VARIATIONS

After you've built a full-sized solar xylophone, it's easy to purchase smaller versions and build modular, portable models. Arrange your portable solar xylophone in a space, and experiment with spatial sound. Weatherproof your xylophones, add hooks or magnets, and attach them to locations in your neighborhood. Add sound anywhere live music is least expected: a garden, outside your window, in sneaky and precarious urban hiding spots.

You can also play around with the orientation and arrangement of the solar cells. With a full-sized xylophone, you can build a truly unique instrument that emphasizes different notes depending on the time of day — an interesting take on the idea of a sun clock. To do this, arrange the solar cells along an arc-shaped bridge, and orient the xylophone so that the arc parallels the sun's path across the sky from east to west.

Want to build more solar instruments? Think about the rudimentary ways that sound is produced — by scratching, hitting, or plucking — and design a solar mechanism that does just that. Use solar cells to automate an already-popular instrument, attempt to re-create an instrument, or build a new instrument of your own.

RESOURCES AND INSPIRATIONS

Solar Beam community site: solarbotics.net

Socrates Sculpture Park, in Long Island City, N.Y., filled with strange sculptures and environmentally powered sound pieces: socratessculpturepark.org

LEMURplex in Brooklyn, N.Y., another great New York institution dedicated to electronic music and interactive art: lemurplex.org

KINETIC REMOTE CONTROL

By Dhananjay V. Gadre

SHAKE THE BATTERIES OUT OF THE PICTURE FOREVER WITH THIS MUSCLE-POWERED INFRARED REMOTE CONTROL.

A TV remote is one of the most commonly used electronic gadgets. We use it without even thinking about it — that is, until the batteries quit.

TV remotes use infrared light for communication with the TV set. Every infrared remote uses AA or AAA batteries to power an infrared LED, controlled by an electronic circuit that beams commands according to the buttons that you press.

This electronic circuit is a very low-power device. Nonetheless, every so often the batteries get drained, usually right when you need them most. And what happens to the used batteries? Perhaps you send them to your local battery recycling plant, or maybe they end up in your town's landfill, polluting the environment. It would be very nice if we could use our remote control devices without any batteries at all!

It turns out that it's easy to bag the batteries, as long as you're willing to put in a little manual effort. How about simply moving your hand back and forth a few times? This motion represents kinetic energy that can be converted into electrical energy, sufficient to power any remote control.

Set up: p.111   **Make it:** p.112   **Use it:** p.115

Dhananjay V. Gadre is an assistant professor with the Electronics and Communication Engineering Division, Netaji Subhas Institute of Technology, New Delhi. Gadre likes to build electronic contraptions for work and for joy. He takes pride in not being associated with any national or international professional cartel, society, or association.

Photograph by Sam Murphy

ZAP THE BATTERIES WITH A FARADAY KINETIC GENERATOR

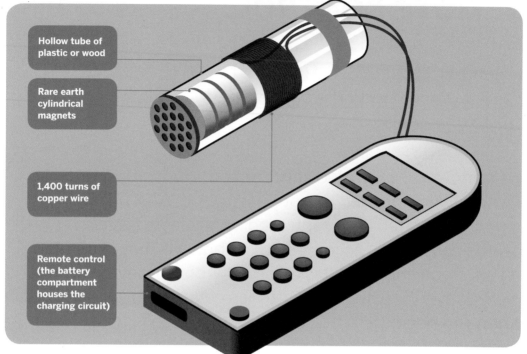

Hollow tube of plastic or wood

Rare earth cylindrical magnets

1,400 turns of copper wire

Remote control (the battery compartment houses the charging circuit)

Infrared remote control devices are available for almost every entertainment gadget at home, be it a TV, DVD player, or music system. Remotes come in all shapes, sizes, and features, including universal remotes that can be taught or programmed to control any given piece of electronic equipment.

Whatever the type, every infrared remote consists of a power source in the form of AA or AAA batteries, an infrared LED, and a set of buttons connected to an electronic circuit that beams a code sequence corresponding to the key that you press.

Unlike most other electronic gadgets, an infrared remote has no On/Off switch. The remote is always on, consuming very little power when in a dormant state. When a button is pressed, the remote goes into an active state, transmits the control code, and then goes dormant again.

This project shows how to retrofit your regular remote control device for battery-free operation, forever. The idea is not so much to save on recurring battery costs, but to remove batteries from the system altogether and, in your own small way, contribute to a greener planet.

The voltage required by the remote control device is generated using a DIY kinetic generator (shown above) that converts mechanical power to electrical power, based on Faraday's principle. The device consists of a hollow tube of plastic or wood, with a cylindrical magnet sealed inside the tube, and 1,400 turns of enameled copper wire wound around the outer surface of the tube.

When the tube is shaken, the magnet travels the length of the tube back and forth. This leads to a change of magnetic field as seen by the wire, and generates an EMF (electromagnetic force) that is proportional to the number of turns and the rate at which the magnetic field is changed. Thus, if you shake vigorously rather than gently, a larger EMF is generated. You can use this principle to easily generate operating voltage for any infrared remote control device.

Illustration by Pars/e Design

SET UP.

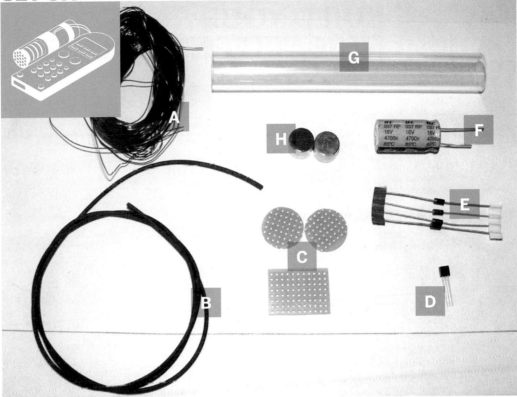

MATERIALS

[A] Enameled copper wire, 30 or 36 gauge, 65' **Enough to roll about 1,400 turns of wire. The gauge of the wire is not critical. I used 36 gauge, but any other, preferably smaller gauge (larger diameter), can also be used.**

[B] 1mm-diameter heat-shrink tubing **to insulate the copper wire going into the remote control. I used Digikey part #A364B-4-ND (digikey.com).**

[C] Perfboard **I used RadioShack part #276-1394.**

[D] Voltage regulator LP2950-3.3 **This is a 3.3V output, low dropout voltage regulator. I used Digikey part #LP2950ACZ-3.3-ND. A 3V output version can also be used.**

[E] 1N5819 Schottky diodes (4) **You can also use 1N5817; I used Digikey part #1N5817-TPCT-ND.**

[F] 4700μF/16V electro-lytic capacitors (2) **Digikey part #565-1538-ND. Two of these capacitors are used in parallel.**

[G] Acrylic tube 6" long, 7/8" OD, 5/8" ID **I used McMaster-Carr part #8532K17 (mcmaster.com).**

[H] Rare earth magnets (4) **cylindrical shape, 1T strength, ½" diameter, Amazing Magnets part #D375D.** amazingmagnets.com

[NOT SHOWN]

Masking tape

A complete kit with all the parts is also available at dvgadre.com/makeremote.

TOOLS

[NOT SHOWN]

Wire snipper
Digital multimeter
Baby hacksaw
TV remote control
Flat file
Soldering iron
Solder
Hot glue gun

Lathe (optional) **to machine the acrylic tube. You can do without.**

Coil-winding machine (optional) **I used one to wind the wire onto the machined section of the acrylic tube. You can also use a power drill (and a friend to help).**

Photography by Dhananjay V. Gadre

MAKE IT.

BUILD YOUR
NO-BATTERY REMOTE

START ⋙   Time: **2 Hours** Complexity: **Easy**

1. BUILD THE FARADAY GENERATOR

The system consists of 4 main components:
1. Infrared remote control device
2. Faraday voltage generator
3. Charge storage component
4. Voltage regulator

Shown here is the schematic of the Faraday voltage generator, the charge storage capacitor (capacitor C1 consisting of two 4700µF/16V capacitors in parallel), and the voltage regulator circuit based on a low dropout (LDO) regulator.

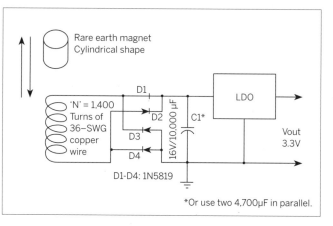

Rare earth magnet
Cylindrical shape

'N' = 1,400 Turns of 36–SWG copper wire

D1
D2
D3
D4

D1-D4: 1N5819

C1* 16V/10,000 µF

LDO

Vout
3.3V

*Or use two 4,700µF in parallel.

1a. Groove the tube (optional). I chose a ⅞" diameter Perspex acrylic tube with ⅝" internal diameter. The length of the tube is approximately 6". You'll need to machine out a 1.5mm-deep groove into the tube, along about 2" of the tube's length. The groove is where you'll wrap the wire. If you don't have access to a lathe with which to cut a groove, the wire can be wound directly onto the plain acrylic tubing. Just use plenty of masking tape to ensure the wire doesn't unwind.

1b. Wind the wire. Fill the groove with 1,400 turns of 36 SWG enameled copper wire. If you have a coil-winding machine, it will be a lot easier. If not, you can improvise with a power drill and a helper: use the drill to slowly rotate the tube while a friend, wearing work gloves, feeds the wire to you. Leave about 6" of wire free on each end. When you're done, cover the wire coil with masking tape, and use 1mm-diameter heat-shrink tubing to cover the 2 free ends of the wire. If you didn't groove the tube, just wind the copper wire directly onto the plain acrylic tube. Be sure to use plenty of masking tape to ensure that the wire doesn't unwind.

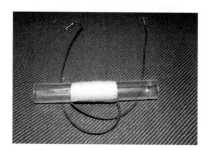

1c. Place the magnets inside the Faraday generator tube, stacking all 4 to make 1 big magnet. Using a hot glue gun, seal the ends with 2 pieces of perfboard, cut and filed in circular shape. To make it more permanent, use a 2-part epoxy.

2. PREPARE THE TV REMOTE

2a. Remove the batteries from the remote control (forever!) and file down the entire battery compartment to remove all extrusions, as shown.

Also file a notch in the side of the battery compartment and the compartment cover, so you can run the wires from the Faraday generator inside.

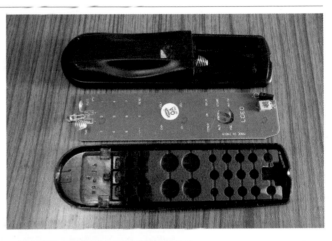

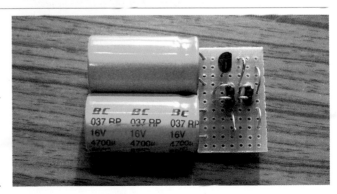

NOTE: You'll need to take the guts of the remote out. Just set them aside.

2b. Assemble the components from the circuit diagram (capacitors, diodes, and LDO voltage regulator) on a perfboard as shown here. You can test the continuity with a multimeter.

2c. Solder the output of the voltage regulator circuit onto 2 hookup wires (extreme right in the photograph) on one end of the perfboard. You'll eventually solder these hookup wires to the battery terminals of the remote control.

2d. Connect the Faraday generator. First solder 2 hookup wires in the middle of the perfboard to the input of the bridge rectifier (cathodes of diodes D3 and D4, respectively, in the circuit diagram). Then solder the wires from the Faraday generator to these hookup wires.

2e. Reassemble the TV remote, and fit the circuit board inside the battery compartment as shown.

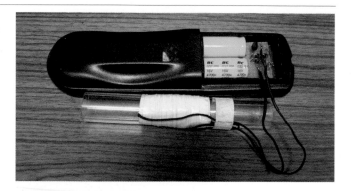

2f. On the remote control's original circuit board, cut down the battery terminals so that they can be soldered to the voltage regulator circuit board. Solder them to your voltage regulator output wires.

2g. Tie the Faraday generator temporarily to the TV remote body using cable ties, and shoot more hot glue between the remote and the Faraday tube. If you want to glue them more permanently, use a 2-part epoxy such as Araldite or J-B Weld's J-B Kwik epoxy. Now you're ready to shake!

FINISH X

NOW GO USE IT »

SHAKE AND SURF

>> To use the device, just shake the remote control back and forth a few times. For my prototype, I shake it about 3–4 times to charge the capacitor, and then I can use it for about 20 key presses.

My first and only "harmonic" (aka my son, Chaitanya) is seen here shaking the battery-less remote to use it to watch his favorite TV program.

If you modify your remote using these instructions, I congratulate you for your effort in preserving the environment. I would love to see the photographs of your completed device! Feel free to mail them to me at dvgadre@gmail.com and post them to the MAKE Flickr pool.

Acknowledgements

I would like to thank Satyaprakash for his ideas and for helping me in this project. I also thank Edward Baafi for his efforts in sourcing many of the components used in this project. I published a different design for a kinetic generator in "Power Generator for Portable Applications," *Circuit Cellar* magazine, October 2006.

Boing Box
By Mark Frauenfelder

Build a fun, one-stringed instrument that packs a mighty twang.

You will need: A cigar box, 6' length of ½"×¾" wood, 8' of 20-gauge wire, 2 eye screws, wood screws, L-bracket, scraps of wood, drill, saw, screwdriver

A 1951 book called *Radio and Television Sound Effects*, by Robert B. Turnbull, shows how to make a "boing box." (It's reprinted at bizarrelabs.com/boing2.htm.) I made a modified boing box using a wooden cigar box and some scraps I had around the house.

1. Drill resonator holes in the cigar box as shown. Screw an L-bracket to the neck, then screw the neck to the cigar box. To prevent structural failure, put a ½"-thick scrap of wood under the lid and drive the screws into it. This will keep the screws from pulling out when the wire is tightened.

2. Insert eye screws into the cigar box and the end of the neck. Use another wood scrap on the inside of the box for the eye screw. Tie and tighten wire to both eye screws. The wire should be tight enough to cause the neck to bow slightly. (I used an eyebolt with hex and wing nuts to make it easy to adjust.)

3. Pluck the string and gently shake the boing box to vary the pitch. Boing!

♪ Listen to some sample twangs of the boing box at makezine. com/12/123_boing.

Mark Frauenfelder is editor-in-chief of MAKE.

Photography by Mark Frauenfelder

THE ATLATL

Make the ancient tool that hurls 6-foot spears at up to 100mph. By Daryl Hrdlicka

Photography by Daryl Hrdlicka

Before the bow and arrow there was the atlatl*, or spear-thrower, an ancient weapon that could throw a spear or dart with enough force to penetrate a mammoth's hide. It was used in North America for about 10,000 years, and used by native Australians and Aleuts as recently as 50 years ago.

It's easy to make your own atlatl, and throwing with it is fun and very satisfying. Here's how to make one in the style of the Kuikuru (kwee-KOO-roo) of the Amazon Basin, who still use the spear-thrower today. I'll also explain how to make darts for it, and how to throw. But never forget that the atlatl is a *weapon*. It is *dangerous*. A dart will go through a side of beef. So I'll go through some precautions as well.

*Most people say "al-LAT-l," or "AHT-laht-l" but pronunciations vary. Find one you like, get your friends to pronounce it the same way, and you'll be right.

Atlatl Basics

Atlatls range in form from the simple to the very ornate, but they all have the same 3 components: the hook, the grip, and the shaft. The grip is where you hold the atlatl, the hook engages the back of your projectile and propels it, and the shaft connects the two and acts as a lever to multiply the speed of your arm.

A typical length for an atlatl is 18"–24", although some have been found as short as 6" (in California) and as long as 48" (in Australia). Length is mostly a matter of personal preference, but it needs to fit the length of your arm and of the dart you're throwing.

The simplest atlatl is the first kind ever used — the basic branch atlatl. To make one of these, just find a tree branch that measures about ¾" in diameter and has a smaller branch angling out of it. Cut it

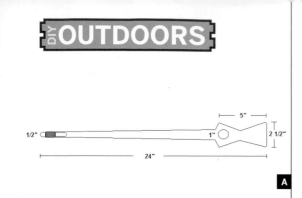

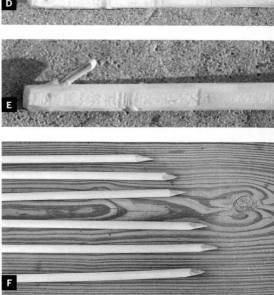

Fig. A: Template for traditional Kuikuru spear-thrower.
Fig. B: Outline traced onto pine board, and ready to
cut. Fig. C: Atlatl shaft cut and sanded smooth.
Fig. D: Drilling the hole for the peg or hook.

Fig. E: Peg installed. If you're using pine (as here),
it's a good idea to reinforce the peg by lashing it with
some cordage. Fig. F: Dart points sharpened with a
hobby knife.

MATERIALS

1×4 lumber, 24" long A standard piece is actually
    ¾"×3½". That's fine.
Wooden peg or dowel, ⁵⁄₁₆"×1¼"
Cordage for wrapping (optional) If you're using pine,
    you'll wrap it. Natural or artificial sinew works
    great, or you could even use kite string.
Wooden dowels, ⅜"×48" one for each dart
Duck feathers, 8" long two for each dart;
    available at most craft stores

TOOLS

Saber saw aka reciprocating saw
Drill with ⁵⁄₁₆" bit, 1" wood boring bit, and ¼" bit
Utility knife
Sandpaper and sanding block
Wood glue
Hot glue and gun for the dart
Strapping or electrical tape for the dart

Making a Kuikuru Spear-Thrower

1. Size and trace the template above (Figure A) onto
your piece of wood, and cut it out (Figure B). Use
the wood boring bit for the finger hole. I normally
make a 1" hole, because I have fairly large fingers
and I share my atlatls a lot. People with smaller
hands can use an atlatl with a 1" hole, but ideally it
should fit close around their index finger; my wife's
atlatl has a ¾" hole. I'd say start with 1", and possibly adjust this for your next one.

2. Use the utility knife to round off the edges, then
sand it all smooth (Figure C).

3. Drill a hole in the end for the peg; this will act as
the hook. Go in at about a 45° angle (Figure D).

4. Put some wood glue in the hole and insert the
peg (Figure E). You're done!

5. If you're using a softwood, reinforce the peg joint
by wrapping it with sinew or cordage. I normally use
pine because it's cheap and easy to shape, but it
tends to break near the peg, so wrapping it helps.

just below the smaller branch, clip off the other end
about 18" farther up, and then clip off the smaller
branch. Now you have a functioning atlatl. To make
it a little easier to handle and control, you can add
a finger loop. Just attach a 10"×¾" strip of soft
leather about 7" from the narrow end, looping it
around on the side opposite the branch stub.

Fig. G: Feathers hot-glued to the tail end of the dart. These slow the tail end down, keeping it in back during flight. Fig. H: First dart finished; now make some more! Fig. I: Larger feathers glued and taped in back so they won't fray as fast. Fig. J: At the Hedoka Knap-In Primitive Skills event, an atlatl thrower steps into his throw. Fig. K: The dart is sent on its way. Note the downward flex in the shaft.

You don't need reinforcement with a hardwood such as oak or maple, but those also require more skill and the proper tools to work effectively — two things I don't have. For your first one, I recommend pine.

Making a Dart

1. Sharpen one end of the dowel (Figure F). I just use a utility knife. You can also add stone, bone, or steel points to darts, but you should probably gain more experience before making your darts lethal.

2. Drill a dimple in the other end. This is the "nock" that the peg fits into. I usually drill with a ¼" bit and just touch the end, going in about ⅛" or so.

3. Hot-glue the feathers on, one on each side, with the quills forward (Figure G). Then wrap tape over the glue to help keep it on, or wrap it with artificial sinew if you want it to look better (Figure H).

4. With large feathers, glue and tape down the trailing ends so that they don't fray as fast (Figure I).

Your dart is done! While you're at it, you might as well make another 5 or so. Otherwise, you'll get very tired chasing it after each throw, and you'll get less practice. Since you're using dowels, your darts will be closely matched, which will help you practice.

Two factors that govern your dart's flight are its flex (also called its *spine*), and the feathers (its *fletching*). Flex is the amount of pressure it takes to make the shaft bend. An atlatl generates 6–10lbs of pressure (depending on your throw), so your darts need a spine of 6–10lbs. Less than that and you won't be able to throw it; more, and it won't fly right.

To measure the spine of a piece of wood or bamboo, press it lengthwise onto a bathroom scale. When the shaft begins to bend, look at the number. That's it. Common ⅜"×48" dowels typically flex in our desired range, as do ½"×72" dowels, which are better for target practice. It's satisfying to make a dart out of a natural piece of wood and hone it to the proper flex, but it's also quite a bit of work. So to start with, I'd say use dowels.

Unlike with an arrow, the feathers on a dart don't act as vanes. They add wind resistance, which slows the rear end so the sharp end stays in front. You can use other materials besides feathers, such as birch bark, cornhusks, cloth, and duct tape, but you can't beat feathers for the look.

For our 48" dowel, a pair of 8" feathers should work fine. With less fletching, the dart will travel farther but won't be as accurate. More fletching means the dart will be more accurate, but won't go as far.

Using the Atlatl

Now let's get out there and throw! The 3 basic steps are the grip, the stance, and the throw itself.

The Grip Slide your index finger through the hole from the side opposite the peg, and grip the handle with your other fingers. Put the point of the dart on the ground, then fit the atlatl peg into the nock and hold the dart with your thumb and index finger. Squeeze them, almost like you're holding a pencil, but keep them on the sides of the dart, not over the back. The dart will come out of your hand at the proper moment, if you just let it.

The Stance Point your left foot at the target (if you're right-handed) and angle your right foot away from it, about a shoulder's-width back. You should feel comfortable and balanced. Turn your body sideways, in line with your left foot, and turn your head to look at the target. Point at the target with your left arm to help with accuracy and balance.

The Throw First, aim the dart by bringing your grip hand up by your ear and sighting along the shaft to your target. Next, bring your arm straight back as far as you comfortably can, but don't twist your wrist on the way back, which will point the dart off to the side. Unless you're a powerful thrower, tip your hand back so that the point rises up a few inches. This will give your throw an arc, making it travel farther. Pause to collect yourself and focus.

And now, the throwing motion itself: using an atlatl is like throwing a fastball — you need to put your whole body into it. Lean back, balancing on your back foot. Then step forward and shift your weight onto your other foot. Slide your arm forward, keeping the dart pointed at the target, and when it's almost fully extended, snap your wrist forward *hard*. It should all be one fluid movement, and the atlatl should end up pointing at the target. For an example, watch the video clips of atlatl throwing on Bob Perkins' website, atlatl.com.

Practice without a dart until you get used to it.

And don't worry about releasing the dart; it should come free on its own at the proper moment. Don't try to throw it hard — this will just mess you up. Just concentrate on throwing smoothly, and your speed and power will develop. Everything will click at some point, and it will be a thing of beauty.

Target Practice

Throwing the atlatl purely for distance is fun, but after a while you get tired of chasing down all your darts. Besides, you'll want to see what it would be like to hunt with one. You can use paper archery targets on hay bales, and 15yds is usually a good starting distance. If you switch to a heavier dart, you'll want to double up the bales.

Standard bull's-eye targets are fine for accuracy competitions, but I personally don't like them. The atlatl is for *hunting*, so I prefer animal silhouettes; 3D targets, your basic foam animals, are my personal favorite. You really feel like the "mighty hunter," and the first time your dart flies straight and true into the target, well, it's indescribable. You need to experience it.

In a pinch, almost anything will make a decent target. A friend and I once used some styrofoam coolers. We ended up hunting those "sheep" for about 3 hours, until it got so dark we couldn't even see them anymore! We were tied at the time (it's always about competition, you know), so we had to keep going, listening to see whether we'd hit them or not. If I remember right, he won. Barely.

⊙ **ATLATL SAFETY — WHERE TO THROW**

» When you throw an atlatl, make sure you have an open area that's at least 30yds long, with nothing breakable behind it.

» There should *never* be anyone in front of you when you throw.

» In spite of your aim, the dart *can and will* go out of control once in a while. It may go off to one side or go farther than you intended. *Make allowances for this.*

Daryl Hrdlicka (thudscave.com/npaa) lives in southwestern Minnesota with his wife and four kids. They homeschool, so making cool stuff is *always* on the menu. Daryl teaches the atlatl at the Jeffers Petroglyphs Historic Site.

SHOOT THE STARS

Astrophotography with your digital SLR camera. By Michael A. Covington

Today's digital SLR cameras can easily photograph more stars in the night sky than you can see — as well as picking up star clusters, nebulae, and galaxies.

Here's how to shoot the stars with just a camera and tripod.

1. Choose your weapons.

A digital SLR (DSLR) camera is best because of its low-noise sensor (Figure A). A point-and-shoot digital camera may work if it has a fast lens (low f-number).

Film cameras work too, if you can get someone to make good prints from the resulting negatives, or if you use slide film so you can see directly what you got.

2. Set up a manual exposure.

Use a 50mm f/1.8 lens if you have one, or set your zoom lens to medium-wide. Put the camera in manual (M) mode, with the aperture at the lowest f-number and the speed at ISO 400. Choose manual focus, too.

If your camera has long-exposure noise reduction (Figure B), turn it on. (On Canons this is deep in the custom settings menu.) This will compensate for "hot pixels" by automatically taking a second exposure just like the first, but with the shutter closed, and then subtracting.

3. Aim at a clear, starry sky.

Put the camera on a tripod and set an exposure time of 5 seconds. Choose delayed shutter release (Figure C) so your finger won't shake the camera. Take aim at the sky, and shoot.

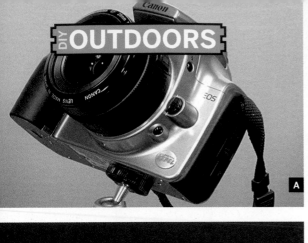

A

B

SHOOTING MENU

Optimize image	✿N
Image quality	RAW+F
Image size	▢
White balance	☀
ISO sensitivity	400
Long exp. NR	ON
High ISO NR	NORM

C

D

Belt of Orion

Rigel (Beta Orionis)

Star cluster NGC 1981

Orion Nebula (M42)

Sword of Orion

Chimney

Fig. A: Look to the skies, DSLRs! This is all you need to do astrophotography. Fig. B: Long-exposure noise reduction automatically subtracts hot pixels. Not all cameras have it.

Fig. C: Delayed shutter release (indicated by arrow) eliminates vibration from your finger pushing the button. Fig. D: Not a bad catch — hundreds of stars, a famous nebula, and a star cluster.

4. Refine the focus.

Your camera can't autofocus on the stars, and the infinity mark probably isn't accurate. So focus manually the best you can, and then take multiple pictures, making small changes in the focus. View each one at maximum magnification on the camera's LCD screen until you've found the best setting.

5. Sharpen and adjust.

Use Photoshop or any picture editing software to sharpen the image (bringing out lots more stars) and adjust the brightness and contrast. The sky background should not be black; more stars are visible if it's only medium-dark.

6. Identify what you've bagged.

Use a star atlas, planetarium software, or online sky atlas (heavens-above.com) to find out what you caught (Figure D). Don't be surprised if your picture shows 10 times as many stars as you could see with the unaided eye.

Michael A. Covington is the author of *Astrophotography for the Amateur* and the new *Digital SLR Astrophotography* (dslrbook.com). By day he does artificial intelligence research at the University of Georgia.

Make: TIPS

Instant Merit Badge:
Animated Knots by Grog (animatedknots.com) shows how to tie a hundred useful knots, with clear animated photo sequences. From necktie Windsors to the Prusik triple sliding hitch (a dangling MacGyver would use one to climb straight up a rope), it's a handy resource for household use, boating, camping, or rigging just about anything with rope or cordage.
—*Keith Hammond*

Spicy Hardware Tip:
A simple spice rack at the workbench can organize smaller bottles of glues, solvents, lubricants, etc. With less than a dollar's worth of hardware they'll mount to the garage wall or pegboard. Variations can be found at hardware stores, home improvement centers, storage retail stores, and some kitchen stores. A simple Google product search for "white wire spice rack" returns various sizes, from a single shelf for less than $3, to a four-shelf version for about $30.   — *Nathan Luoto*

Find more tools-n-tips at makezine.com/tnt.

This well-used kids' BMX bike needed only new tubes and some chain oil to return it to use.

BIKE SCROUNGING

How to fix a castoff bike and give it away.
By Thomas Arey

Photography by Thomas Arey

I'm going to venture a guess that many makers' earliest experiences working with tools and trying to figure out machines involved a bicycle. Even today it's the rare kid who hasn't tried to fix or even modify their bike. It's one of the reasons I still have great hope for humanity.

Cycling is good basic transportation, a boon to the cardiovascular system, and most of all, fun! But have you ever considered that cycling can also be free?

In the course of the trash picking and dumpster diving I do to bring these occasional articles to MAKE, I often run across bicycles left at the curb with other signs of our society's tendency to toss away what might be repaired or repurposed.

I've taken many of these rejected rides, turned them back into working bicycles, and donated them, either locally or through service organizations, to folks whose lives can literally be changed by owning a bicycle.

My general experience shows that the parts from 2 or 3 disposed bikes can make for 1 good bike. Any leftover parts from each scrounging venture go into storage to support future bike recovery operations.

Bicycle recovery is the perfect "learn by doing" process. Beyond stripping some threads (also repairable) you can't really hurt anything. Mixing and matching parts from different bikes will make you more adept at repair. This can even turn into a marketable skill with enough practice. Good bicycle mechanics are hard to find.

Your public library and the internet will turn up dozens of books and websites to help you get beyond the basics quickly. A good book that covers just about everything you need to know and more is *The*

Bicycling Guide to Complete Bicycle Maintenance and Repair: For Road and Mountain Bikes by Todd Downs.

Sheldon Brown (sheldonbrown.com) is a well-known cycle mechanic who shares tons of information free on the web. A good source for odd and hard-to-find parts is Loose Screws (loosescrews.com).

Older bikes and most consumer-grade cycles can be worked on with common hand tools. The only specialty tool you may need from the start is a chain tool, required to remove and replace the chain on most multigear bikes. This tool can be found for as little as $10, but if you plan to do this a lot you should invest in high-quality tools.

When you come upon a bike leaning against a trash can, don't assume it's being trashed. I always knock on the door and check. More often than not I hear, "I got a couple more around back, you want them, too?"

When you get your bike(s) home, go through these steps:

1. Check your find over. Why was this bike tossed? I am always surprised to find that a few small problems led to the trip to the curb: a flat tire, snapped brake cable, or rusted chain being the most common.

2. Once you fix up the obvious problems, go over every nut, bolt, and bearing to tighten things up and check for more subtle problems that may require further disassembly.

3. In most cases, if it moves, lubricate it! Extremely neglected bikes may require greasing the bearings, but a little chain oil will get most bikes back on the road.

4. Replace bad or worn parts with other items from your trash-picking efforts. Get friendly with your local bike shop. They have trash bins, too!

5. Even if the tires inflate, check both tires and tubes for signs of dry rot. Well-cared-for tires can last a long time but this may be the one place you need to spend money.

6. Broken spokes and bent wheels are intermediate-level repairs. Until you master the skills for this task just keep an eye out for other good wheels on your scrounging route.

Fig. A: An old ten-speed with a rusty chain and a few missing parts. A perfect project bike for personal use or for donation. Fig. B: A reincarnated Raleigh road bike ready to ride for many miles.

7. Double-check all matters of safety, especially the braking system.

8. Enjoy the ride. It may be a little rough and rusty, but it rolls and the price is right!

After you've built a bike or two for your personal needs, why not think of getting your rebuilds into the hands of folks who can use them? Check your local social-service and faith-based organizations.

If you want your bikes to go beyond your local neighborhood to help the world, one clearinghouse website for bike donation is the International Bike Fund's page at ibike.org/encouragement/freebike.htm. This site lists organizations throughout the United States and other countries, and includes details about how your efforts to repair and reincarnate castaway cycles can truly work to change the world.

T.J. "Skip" Arey N2EI has been a freelance writer to the radio/electronics hobby world for over 25 years and is the author of *Radio Monitoring: The How-To Guide.*

CHASING THE GODSHOT

Pack perfect grounds every time with this hydraulic espresso tamper. By John Edgar Park

Photography by John Edgar Park

Godshot. That's the elusive goal of espresso fanatics everywhere. Thick with micro-bubble crema and a velvety mouth-feel, and packing explosive flavor, the godshot is pulled too infrequently for my liking. A perfect shot of espresso is the product of many variables, so anything I can do to lock in one of those variables is a good thing. Think scientific method as applied to espresso.

As a home barista, the five factors I worry about most are: beans, water, grind, dosing, and tamping. Here's a quick overview:

Beans: Use high-quality beans that have been roasted within the past 3 days to 2 weeks.

Water: Fresh water needs to be brought up to proper temperature and pressure by the espresso machine.

Grind: Dial in your burr grinder to fine-tune the coarseness of the ground beans.

Dosing: There's a prescribed amount of grounds to measure into a standard portafilter basket, but people tinker with this all the time.

Tamping: Once the grounds have been measured into the portafilter and distributed evenly, they should be compressed with 30 pounds of pressure using a tamper. This is difficult to do by hand.

Once these steps have been taken, the portafilter is locked into the group head of the espresso machine and the shot is pulled. This is sometimes accompanied by breath holding, finger crossing, and the like.

The variable I chose to attack was the problem of tamping. I'd heard about automated tamping systems, which remove the guesswork of manually pressing grounds into the portafilter with a hand-held tamper, and instead consistently provide 30 pounds of uniform pressure. This is great, but automatic tamping machines cost hundreds of dollars. One clever member of the home-barista.com forums scoffed at the price tag and built his own, using a manual lever juicer as the platform. I decided to give it a try.

I considered a few requirements for my auto tamper: a way to know when I've hit the magic 30lbs, a way to secure a commercial tamper base to the down-shaft of the juicer, and a method of resting different portafilters with their oddly shaped undersides on a load-bearing surface.

Through some informal research, I determined that the most common method of indicating correct pressure in a commercial auto tamper is with a "clicker" system. This is a calibrated spring and ball bearing mounted at a right angle to the shaft. One great feature of these systems is that they stop all downward pressure after the threshold has been met.

Home baristas have more commonly opted for a heavy spring mounted over the juicer shaft, kind of like a car's coil-over suspension. Enlisting the aid of a bathroom scale, they test for the 30lb depth and then mark a calibration line on the shaft. This is an elegant, simple design, and they've reported excellent results. Another simple method, suggested by a gearhead friend of mine, would be to replace the lever handle on the juicer with a torque wrench adjusted to pop when 30lbs are measured at the tamper.

While the torque wrench would have functioned well, I was also concerned with aesthetics. I wanted to maintain the quasi-steampunk look of my espresso machine, with its beautiful pressure gauges front-and-center. I wondered how I could use an analog dial to read out my tamping force. After another brainstorming session with my friend, we had it: a pressure gauge connected to a hydraulic piston coupled between the juicer's down-shaft and the tamper base.

To set these plans into motion, I began sourcing parts. The foundation for the project is the juicer. It can be had for about $20 at T.J. Maxx or Marshalls, and I also found many on eBay — just search for "manual juicer."

The piston, gauge, and fittings proved a bit harder

MATERIALS

Manual lever juicer around $20 from department store clearance sales or ebay.com
Espresso tamper base You may be able to unscrew the handle of your current tamper, or buy one at coffeetamper.com.
Actuator piston See what you can find at the salvage yard, or search eBay for a 1" Parker actuator.
100psi pressure gauge
Aluminum 1" round stock, 1½" length Try a local metal supply or order from Small Parts (smallparts.com).
Aluminum ½"×2½" flat bar stock, 5" length
Fittings and pipe to connect piston to gauge
Glycerin or castor oil Either makes good hydraulic fluid, and neither should accidentally poison you!
Threaded insert
Thumbscrew or bolt

Total cost should be $60–$100, depending on where you find your parts.

TOOLS

Needlenose pliers to remove lever handle retention clip
Crescent wrench
Drill press and various bits
Tap wrench and taps
Hex wrenches for juicer disassembly
Angle grinder to cut and smooth aluminum

to track down. I found many sources online, but they were all too expensive. I finally hit the jackpot at Norton Sales (nortonsalesinc.com), an aerospace scrap yard in North Hollywood, Calif. There, I picked up a stainless steel actuator piston for $5, a 100psi gauge for $7, and the pipe and fittings for a few bucks.

I went to Industrial Metal Supply (imsmetals.com) for scrap aluminum stock. A neat little store in

Disassembling the juicer. First, remove the cone-shaped cup from the juicer shaft (save it — it may come in handy for a future project). Figs. A and B: Remove the lever assembly.

Fig. C: Use pliers to remove the retention clip on the lever handle. Fig. D: Place the down-shaft at its highest point, and reinsert the lever handle at the starting angle, as shown here.

Connecting the juicer shaft to the actuator piston. Fig. E: Cut a 1½" length of 1" aluminum rod. Fig. F: Drill and tap the rod to couple the juicer shaft

and the piston mounting thread. Figs. G and H: Juicer shaft, coupler, and piston shown separately and joined together.

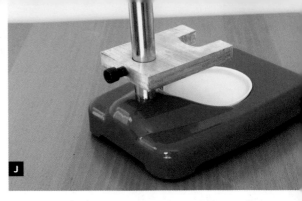

I

J

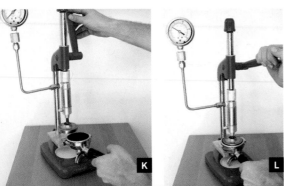

K

L

M

Final steps. Fig. I: Thread insert into tamper base to join it to the actuator piston. Fig. J: Make base plate from rectangular aluminum stock. Cut a U-shaped notch to hold the portafilter and add a bolt to tighten the base onto the support column. Figs. K–M: Pull the lever to tamp your grounds, while keeping an eye on the pressure gauge. Perfect tamping!

Burbank, Calif., called Luky's Hardware yielded various bolts, inserts, taps, and cheap, resharpened drill bits.

To begin construction, I disassembled the juicer and its down-shaft. I then removed the lever handle, placed the down-shaft at its highest point, and re-inserted the lever handle at a better starting angle.

Next, I drilled and tapped a 1½" length of 1" aluminum rod to couple together the juicer's shaft and the piston's mounting thread. If you don't want to drill and tap solid bar stock, you could consider using a tube with some cross-drilled tightening bolts instead.

I couldn't find an insert of the correct dimensions to mate the piston rod to my commercial tamper base. I found one that was very close, and cut it with a die to fit the coarse threading of my tamper. A bit of a hack, but it seems to work well enough. If you're willing to re-tap your tamper base, you'll have a wider variety of inserts to choose from.

I filled the piston with glycerin, my food-safe hydraulic fluid of choice, and then screwed in all the fittings for the pipe, gauge, and piston (including a plug for the upstroke inlet). I left a bit of air in the line to provide a little compression. I found that it feels better to have some give as you pull down on the lever. If this were a car, you'd call it mushy brakes.

Since I use both regular and bottomless portafilters (see MAKE, Volume 04, page 117, "The Bottomless Portafilter" to build your own), I needed a versatile enough base plate to support either of them during tamping. I cut a 5" length of my rectangular aluminum stock with a grinder, and then cut a U-shaped section out of it. I made this wide enough to clear the underside of a regular portafilter but still support the bottomless one.

I drilled a ¾" hole at one end of the plate so I could slide it over the juicer's support column. Finally, I drilled and tapped a ⅜" hole at the back of the plate. This allows me to tighten it in place with a small bolt.

With everything assembled, I've now got my own auto tamper for about $50 — far less than a commercial unit. It functions flawlessly, gives me an exact tamp every time, and looks stylish. But most importantly, it's removed tamping from my list of espresso worries in the continued quest for the godshot.

John Edgar Park rigs CG characters at Walt Disney Animation Studios. His home base is jpixl.net.

Make:
technology on your time™

Sign up now to receive a full year of **MAKE** (four quarterly issues) for just $34.95!*
Save over 40% off the newsstand price.

NAME

ADDRESS

CITY STATE ZIP

E-MAIL ADDRESS

MAKE will only use your e-mail address to contact you regarding MAKE and other O'Reilly Media products and services. You may opt out at any time.

*$34.95 includes US delivery. For Canada please add $5, for all other countries add $15.

makezine.com/subscribe
For faster service, subscribe online

promo code **B7DTA**

makezine.com

Make:
technology on your time™

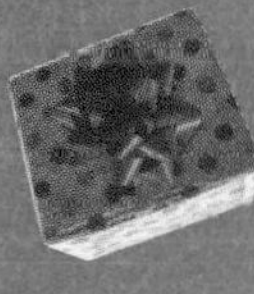

Give the gift of MAKE!

makezine.com/gift
use promo code **4DGIFT**

When you order today, we'll send your favorite Maker a full year of MAKE (4 issues) and a card announcing your gift—all for only $34.95!*

Gift from:

Name

Address

City   State

Zip/Postal Code   Country

email address

Gift for:

Name

Address

City   State

Zip/Postal Code   Country

email address

*$34.95 includes US delivery. Please add $5 for Canada and $15 for all other countries.

47GIFT

NO POSTAGE
NECESSARY
IF MAILED
IN THE
UNITED STATES

BUSINESS REPLY MAIL

FIRST-CLASS MAIL   PERMIT NO 865   NORTH HOLLYWOOD CA

POSTAGE WILL BE PAID BY ADDRESSEE

Make:

PO BOX 17046
NORTH HOLLYWOOD CA 91615-9588

NO POSTAGE
NECESSARY
IF MAILED
IN THE
UNITED STATES

BUSINESS REPLY MAIL

FIRST-CLASS MAIL   PERMIT NO 865   NORTH HOLLYWOOD CA

POSTAGE WILL BE PAID BY ADDRESSEE

Make:

PO BOX 17046
NORTH HOLLYWOOD CA 91615-9588

BOOMBOX AS PLATFORM

This cheap garage-sale stalwart is really an electronics lab in a box. By Mister Jalopy

A MORE ROBUST $1 PLATFORM DOES NOT EXIST!

LOOK FOR:
- » Durable plastic enclosure
- » Ample interior space for mods
- » Carrying handle
- » AM/FM radio
- » Antenna
- » Stereo cassette deck
- » Audio amplifier
- » AC to DC converter
- » Battery box
- » VU meter
- » RCA line level inputs
- » RCA line level outputs
- » Built-in stereo microphones
- » 2-way stereo speaker system
- » Microphone input jacks
- » Headphone jack
- » External speaker jacks
- » 110/220 power selection
- » AC or battery-powered
- » Made in Japan

The Good Old Days Are Yet To Come

Being a collector of old issues of *Popular Mechanics*, *Mechanix Illustrated*, *Modern Mechanics and Invention*, *Popular Science*, and *Science and Mechanics* magazines, I can attest to the awesome breadth of the handy heydays when every home garage had at least a modest workshop. But with MAKE magazine as our beacon, I am quite certain that we have not yet seen the Maker Golden Age. As the internet has connected like-minded individuals and the availability of information has exploded, the real catalyst in the maker movement is the staggering abundance of dirt-cheap high technology.

Before boomboxes became cheap Chinese commodities, the top-tier Japanese electronics companies would compete to offer the most features, the best sound, and the highest profit margin, as reflected in

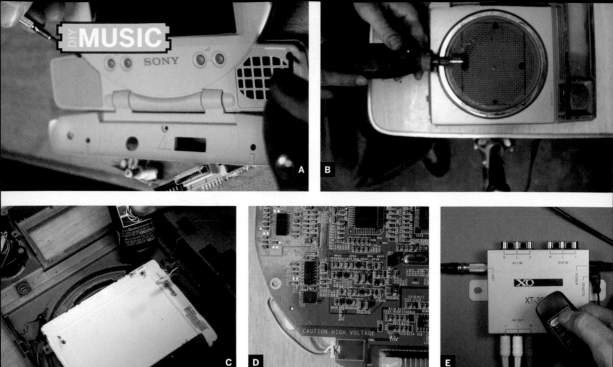

Fig. A: The Sony PSone screen is a great, hackable display perfect for installing in a boombox. Fig. B: Use a Dremel tool to cut out a rectangle in the speaker grill to accommodate the display. Fig. C: Use RTV black silicone from the auto parts store as a glue to attach the display to the boombox. Fig. D: Heed all warning signs regarding electricity! Fig. E: An RF tuner provides a television signal.

the lofty retail prices enjoyed at the time. Now, they are invariably under $5 at garage sales, with the earlier cassette (as opposed to CD) models being the more robustly optioned units. Having a shelf of boomboxes, I use them for all sorts of projects, including the audio system for my Urban Guerilla Movie House (see MAKE, Volume 11, page 48).

My first boombox hack was to add a TV. I picked up a Sony PSone screen (an option for the first-generation PlayStation). This is a great, hackable NTSC screen available on eBay for a fraction of its original cost. HINT: The case screws are hidden under the speaker grills (Figure A).

I made an LCD-screen-sized cardboard template and sacrificed one speaker. When cutting soft plastic, go slow with the Dremel cut-off disc as plastic would just as soon melt as be cut (Figure B).

When mounting loosey-goosey, depackaged components, I often use RTV (room-temperature vulcanizing) black silicone from the auto parts store as a glue (Figure C). Its attributes are so great that I should write a sonnet about my RTV devotion. It is nonconductive, so it's a great insulator. It's also a great shock mount, as it dries like a piece of cast rubber. In addition to being a swell adhesive, you can use it like caulk to fill gaps and mistakes!

⚠ **DANGER:** By the very nature of case modding, you are removing safety measures, electrical shielding, and engineered protections (Figure D). This is no place to be cavalier! Be sure that electrical connections remain shielded and that the final product is sealed up safely.

The secret weapon of the TV Boombox is the XO Vision XT-3000 TV tuner (about $70). Made for the car video market, XO is a flexible, compact, and elegant RF tuner solution. It's got 2 AV inputs, remote control, IR remote receiver, and antenna input, and it's so energy efficient that I was able to power it with a 12V DC 300mA cellphone power adapter (Figure E).

Make: TIPS!

Broadcast FM from Your Stereo or iPod:
For an attractive and inspired distributed sound system, sprinkle boomboxes all over your estate and use a cheap consumer transmitter.   —Mr. J

Mister Jalopy is a mediocre welder, a fair shade-tree mechanic, and a clumsy designer, and has never touched a piece of wood he hasn't ruined. However, he still gets a lot of love at hooptyrides.com.

STYROFOAM PLATE SPEAKER

Get surprisingly good sound from disposable picnicware. By José Pino

I've built homemade speakers using various materials for the cone. This design is the best. Paper plates are too soft, and disposable plastic cups vibrate too much, but stiff, lightweight styrofoam produces sound quality that competes with commercial speakers. I really mean it — you will be surprised!

1. Cut 2 strips of paper, ½"×11" each. Coil one strip lengthwise around the magnet and tape it closed, but don't tape the paper to the magnet. Roll the other strip around the first one and tape it closed as well, but don't tape it to the first roll. Remove the magnet (Figure A).

2. Glue the paper cylinders to the back of the plate. Try to position them at the exact center (Figure B).

Photograph by Sam Murphy

MATERIALS
Foam plate
Sheet of regular bond paper
Business cards (2)
Copper wire, 32-gauge enameled
Tape
Glue **Hot glue works great.**
Small neodymium cylinder magnet(s) **You can use just 1 if it's tall enough, but I used 3 thin ones stacked together.**
Flat piece of wood or cardboard **Should be larger than the plate. I used cardboard, but wood damps vibrations better.**
Audio plug

TOOLS
Ruler
Scissors
Wire cutter/stripper

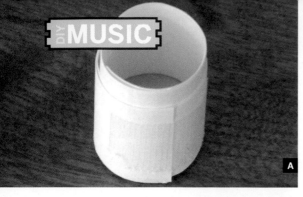

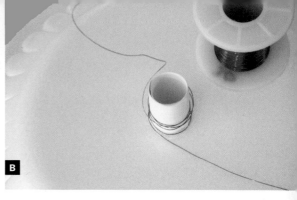

Fig. A: Coil paper strips lengthwise around the magnet, taping each one closed (but not to each other). Then remove the magnet. Fig. B: Wind 50 turns of wire around the paper coil and secure with tape.

Fig. C: Put the magnet in the cylinder and glue the magnet to the cardboard base and the folded business cards to the plate. Fig. D: Cut and strip the wires from the audio plug and connect them to the coil.

3. Put the magnet back inside the cylinders, and then wind the wire around them, about 50 turns. The coil should have more than 7 ohms of impedance. Leave some extra wire length at each end, and after you finish winding, secure the coil with tape (Figure B).

4. Remove the magnet and then remove the inner cylinder. It's OK to tear it, but try not to damage the outer cylinder.

5. Accordion-fold the 2 business cards widthwise into W shapes. Then glue 1 end of each to the back of the plate, so the cards are parallel and stick up symmetrically on either side of the cylinder.

6. Replace the magnet in the cylinder. Put some glue on the free ends of each business card, and a little bit on the magnet, but not enough to squeeze away and touch the inside of the cylinder. Position your piece of cardboard or wood on top, with the plate centered underneath — this will be the speaker's base. Flip everything over so the base is on the bottom and the plate faces up. The magnet should fall down and glue itself onto the base, with the business cards glued symmetrically on either side. The cylinder should hover around the top of the magnet (Figure C).

7. Make sure that the coil wires are separated and aren't touching anything. Cut and strip the wires from the audio plug and connect them to the coil (Figure D). Allow glue to dry.

Your homemade speaker is ready! I plugged mine into my computer. The volume was good, and the sound quality was really good. I can listen to music across the room.

Troubleshooting

Make sure the business cards are parallel, and try gluing them closer to or farther away from the coil, until you find the distance that produces the most and best sound.

If your speaker sounds horrible, check that the coil is tight and secure, with nothing loose, and that all other wires move freely, without anything touching them. Make sure the business cards are completely glued and that they're the only things that touch the foam plate. The coil should not touch the magnet or the base of the speaker. If it does, make the coil wider or don't fold the cards as tightly.

José Pino is an inventor who lives in México and likes electronics. See his projects, crazy ideas, and more at josepino.com.

Photography by José Pino

DIY
HOME

$1,000 WIND CHIMES

Finding the softer side of hard drive platters. By Thomas Arey

Over the years, I have accumulated a number of computer hard drives which either failed or needed upgrading. These old drives were no longer collecting data; they were collecting dust. So it was time to void the warranties and open the cases.

Fair warning: Cracking open any hard drive will probably stop it from working. These units are assembled in clean-room conditions, and while my basement workshop is a fantastic place full of wonders, it has never been described as clean. I knew that whatever came out of this exercise would have to perform some new function.

Most modern hard drives are assembled using small Torx screws. With a well-stocked toolbox this is no problem, but before you start unscrewing, notice the thick metallic tape that runs around the drive to join the 2 halves of the drive case and seal the innards against the outside world. A quick slice with a razor knife takes this encumbrance away and lets the unscrewed case come apart. Keep an eye out for screws hidden under labels.

Inside we find the drive head assembly, a small printed circuit board, and one or more disk drive platters sitting on a small motor. The most generally useful, recoverable item inside any old disk drive is the powerful neodymium magnet that pulls the drive head. At the very least, these make geeky refrigerator magnets. I have joined several of these magnets together and threaded strong cord through their plates to go "fishing" in a lake. They won't pull up a car wheel, but you may find some interesting underwater debris.

With that said, I began to think a little bit more artistically about the shiny platters. I gathered

Photography by Sam Murphy

A B

C D E

Fig. A: Clamp platter with a rag to protect its surface, and use a cutting wheel to free extraneous pieces.
Fig. B: Smooth, shiny platter with pieces removed from its center.

Fig. C: Measure aluminum strip for hanger ring.
Fig. D: Mark pilot holes for drilling hanger ring.
Fig. E: Rivet (or bolt) ends of ring together.

MATERIALS

Old hard drives
1" strap aluminum or other material to make a ring
Small nut and bolt or pop rivet tool and rivets
Monofilament fishing line or thicker cord to avoid tangling with longer hanging distances
Torx screwdrivers
Knife
Drill

several together and spread them out on my workbench. As I pondered their mirrored surfaces, I heard one of my wife's wind chimes sing from our back deck. In a magic maker moment, an LED lit up in my head. Let's turn these platters into wind chimes!

In my scrap metal pile I found some 1" strap aluminum that would make a nice hanger ring. I laid my 7 platters out in a line on my workbench, with their edges overlapping by about ¼", and then took this total measurement. After adding about 1" for overlap, I had the length of aluminum I needed.

I drilled a hole at each end of the aluminum, for joining the ring together using a small nut and bolt. Then I drilled 7 evenly spaced holes for the platters to hang from. Someone more math-oriented than myself could come up with a formula to compute

the hole placement, but I found that eyeballing got me to where I wanted to go. Besides, it was fun.

At first I tried hanging all the platters at the same height below the metal ring. I strung clear fishing line and hung them down about 16". These "first draft" wind chimes sounded nice, and the sunlight reflecting off the platters bathed my backyard in "fireflies" that my dogs had fun chasing. However, strong winds led this design to tangle very quickly and stay tangled. Back to the workbench!

Looking at other wind chimes around my house, I noticed that many had their bells arranged in a cascade of successively longer strings. When these tangled in strong winds, they tended to untangle on their own. Making this simple change in the lengths of fishing line made for an even more attractive kinetic sculpture with a fine sound.

After I hung these chimes, several neighbors asked if I could build some for them. Since the original cost of the hard drives exceeds $1,000, folks will just have to wait until I can find more dead ones.

T.J. "Skip" Arey N2EI has been a freelance writer to the radio/electronics hobby world for more than 25 years, and is the author of *Radio Monitoring: The How-To Guide.*

LEGO RECHARGER

It's a snap to keep your gadgets juiced and your keys from getting lost. By John Edgar Park

Photography by John Edgar Park

On a recent trip to Legoland, I saw a neat product in one of the stores: a Lego key rack with Lego brick keychains. What a great idea, I thought. With this I could come home, empty my pockets, and have a consistent place to hang my keys. But wait, what about all the other devices I just pulled out of my pockets, where do they go? And, for that matter, how will all their batteries stay charged?

Then it dawned on me. If I attached a powered Lego brick to each gadget to provide life-giving juice for their thirsty batteries, I'd solve 3 major problems in my life: lack of gadget organization, lack of battery power, and lack of Legos attached to all my possessions.

The first thing I did was to sift through my Lego Mindstorms and Technic bins. I grabbed some 9V motor wire bricks and a large baseplate to start

playing with the design. I wanted to avoid modifying the bricks as much as possible. I also wanted color coding so I'd be less likely to accidentally hang my iPod on the cellphone's brick, thus blowing up the iPod. This is a danger of universal connectors. Since the motor wire bricks come only in black, I needed to use additional bricks for color coding. I considered color-coded tiles on top of the device-end brick, but the smooth tiles always seem to hide a Lego's, well, Lego-ness, so I opted for a 2×2 studded plate instead. Much more geek chic. I placed color-matched bricks below the respective charger-side brick on the base plate.

Next, I needed to splice the motor wire bricks onto my power adapters and gadget plugs. My first attempt involved cutting, stripping, and twisting corresponding wires together, soldering them, and

MATERIALS

Lego 8×16 baseplate brick part #4204
Lego electric bricks 2×2×²/₃ with 9V wire end (2)
    part #5306 **aka motor wire**
Lego 2×2 bricks (8) part #3003
Lego 2×4 bricks (2) part #3001
Lego 2×2 plates (2) part #3022
#214 screw eyes (2)
#6×1½" wood screws (4)
Various DC power chargers and devices

You can find most of these parts at online Lego
stores on bricklink.com. Better yet, dig through
your Lego bins and adapt the design to suit what
you've got on hand.

TOOLS

Needlenose pliers
Small flathead screwdriver
Wire strippers
Diagonal cutters
Multimeter
Dremel tool with small drill and burr bits
    for countersinking wood screws
Soldering iron **(optional)**
"Third hand" tool

A

Fig. A: Pry off the back panel of the Lego 9V motor
wire bricks using a small screwdriver. Discard or save
the wire, as you won't need it for this project.

I'd now be able to cut all of my charger wires in
half, and simply crimp a Lego motor wire brick
onto each end.

For my key chain, I ripped off the original Lego
design. I drilled a small hole into a 2×4 brick and
then screwed a small screw eye into it. My apologies
to Lego purists for all the drilling, but hey, Lego
did it first!

The whole system was cheap and easy to build,
works great, and keeps my devices organized and
charged. I've gotten so used to it that I've installed
an unwired counterpart key rack at my office.

1. Splice the gadget chargers with Lego brick connectors.

1a. Pry off the bottoms of both bricks of a Lego
9V motor wire for each gadget you plan to adapt.
Remove the black Lego wire; you won't be using
it (Figure A).

1b. Plug in your charger, and test the plug with a
multimeter to determine voltage and ground. For
example, my Nokia phone charger has a negative
exterior and a positive interior. Make a note of this,
so you can double-check your work later before
plugging the device in.

1c. For devices that carry data as well as power
over their cables, such as iPods, you might need
to consult an online wiring diagram to determine
which pins carry voltage and ground.

1d. Unplug the charger, then snip the charger wire
¾" from the device-end plug. Carefully remove ½"
of the outer tubing to expose the insulated inner

then covering the splice with heat-shrink tubing.
This worked great, but wasn't very elegant. I wanted
to leave these Lego dongles on my gadgets all the
time, even when they were in my pockets, so getting
the wire length down to a minimum was important.
The splice wasn't helping that.

Looking more closely at the Lego 9V motor
wire brick, I noticed 4 pressure tabs on its ends.
I grabbed a small screwdriver and pried the bottom
off the brick. Inside, the insulated wire pair was
pierced onto 2 sharp metal posts. The wire was held
in place by the pressure between a small ridge of
plastic and the recently pried-off bottom. Excellent.

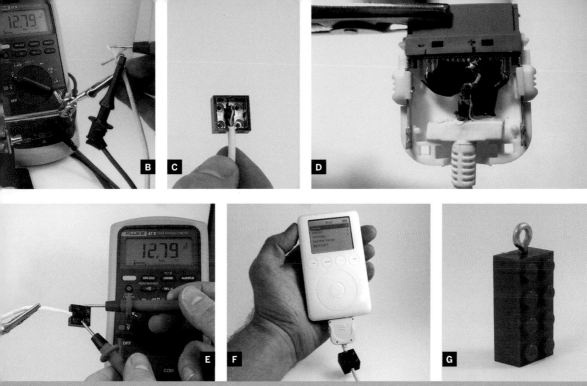

wires of the charger and plug. Using a multimeter, determine the + and - wires on the charger and plug. Label these if they aren't color-coded, so you can make the proper connections to the bricks (Figure B).

1e. Lay the 2 power wires from the charger over the metal spikes of the Lego brick and push them into place with a small flathead screwdriver. Push hard enough to pierce the insulation. For extra insurance you can solder them in place (Figure C).

1f. Repeat Step 1e for the device plug wires. Note which spike carries the positive charge, and remember to keep the circuit intact by connecting the positive wire to the proper spike on the plug-side brick.

For my 3rd-generation iPod dock connector, I couldn't get voltage to the pins after splicing the wires. Once I finished cursing at Apple for using such ridiculously tiny connector pins, I fixed the problem with a $4 unsoldered dock connector from SparkFun Electronics (sparkfun.com, part #DEV-00633), by soldering the ground to pin 29 and the 12V DC+ to pin 19 (Figure D).

It's worth noting that I killed a battery by grounding to pin 1 initially. At least I didn't destroy my iPod.

1g. Push the bottoms back onto the bricks, listening for the satisfying "snick" of the tabs popping back into place. If it's overly difficult to snap on, try moving a wire to one side of the plastic ridge inside the brick. A single pinched wire will hold the weight of most pocket-sized gadgets.

1h. Check your work by connecting the bricks with proper orientation, plugging in the charger, and testing voltage/polarity with a multimeter. This needs to read the same as it did in Step 1b, or else you may fry your costly gadget. Once you're satisfied that it's all wired correctly, plug your device into its Lego plug to test that it charges (Figures E and F).

2. Create the Lego keychains.

2a. Drill a hole with the Dremel into one end of a 2×4 brick, passing through one cylinder wall on the underside.

2b. Screw in a screw eye, then attach your keys (Figure G). Assuming you already had the ubiquitous 2×4 brick laying around, but had to buy the screw eye, this step just saved you about $3.98 off Lego's $3.99 price for the same thing.

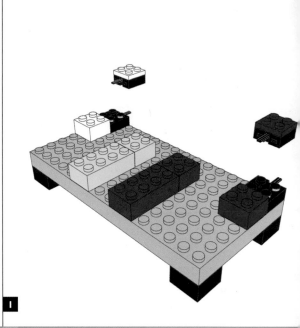

H **I**

Fig. H: The white, yellow, red, and blue bricks shown here are used as a color-coded gadget identifier.
Fig. I: Build the Lego charger base station by running charger brick wires over the top of the base brick (shown here in gray) and running them behind the base brick. You can use a Lego software application, such as the free MLCad (ldraw.org) to try out different designs in advance.

J **K** **L**

M **N** **O**

Fig. J: Color-code the connections with a 2×2 plate on the device end and a 2×2 brick on the charger end.
Fig. K: The 2×2 brick below the power wire brick will help prevent your gadget from falling. Fig. L: Drill holes into four 2×2 bricks and add screws. These "screw bricks" will be used for wall-mounting. Fig. M: Snap screw bricks onto base brick. Measure, mark, and drill holes in wall. Figs. N and O: Mount charger on wall.

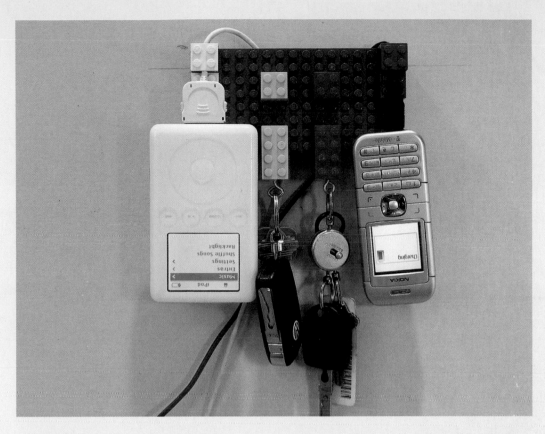

3. Build the Lego base station.

3a. Follow these illustrations to build the base station. Run the charger brick wires over the top of the 8×16 base brick and then behind it (Figures H and I).

3b. Color-code the connections with a 2×2 plate on the device end and a 2×2 brick on the charger end. I used a white plate on my iPod plug, and a white brick under its corresponding charger brick. Since the wire bricks are ⅔ normal brick height, the 2×2 brick provides a nice little ledge for additional insurance against your gadget falling tragically to the floor (Figures J and K).

3c. Using the Dremel, carefully drill a hole for a wood screw into 4 of the 2×2 bricks. Use a burr bit to create an angled hole so that the screws can be countersunk. Now drop the wood screws into the holes (Figure L).

3d. Mount the base station onto the 4 screw bricks. Measure, mark, and drill holes for mounting on your wall. Make sure you're near enough to an outlet to plug in the chargers. Depending on the wall you'll be mounting this on, you may need to drill the wall for anchors as well (Figure M).

3e. Pull the 4 screw bricks from the base station, then screw them into the wall. Remount the base station to the wall (Figures N and O).

3f. Plug in your wall chargers, hang your keys and gadgets on their new home, and, most importantly, bask in the immense inner calm of knowing your stuff is now neatly organized and humming with power (pictured above).

John Edgar Park (jp@jpixl.net) is a character mechanic at Walt Disney Animation Studios.

Small-Order Metal:
Speedy Metals (speedymetals.com) is a great place for ordering small quantities of metals for your projects. I've ordered brass tube stock, 7075-T6 aircraft-grade aluminum, and tool steel from them. Shipping is fast, and they are geared toward small orders — so they're not "put out" when you only order only one or two items.

—Devon Prescott

Find more tools-n-tips at makezine.com/tnt.

BEETLEBOT

Ultra-simple bugbot navigates obstacles with feelers and switches. By Jérôme Demers

The Beetlebot is a very simple little robot that avoids obstacles on the floor without using any silicon chip — not even an op-amp, and certainly nothing programmable. Two motors propel the bugbot forward, and when one of its feelers hits an obstacle, the bot reverses its opposite motor to rotate around and avoid it. The project uses only 2 switches, 2 motors, and 1 battery holder, and it costs less than $10 in materials (or free, with some scrounging).

Beetlebot in 10 Easy Steps

1. Cut pieces of heat-shrink tubing and use a heat gun or other high-heat source to shrink them onto the motor shafts. Trim the tubing evenly, with a little bit running past the ends of the shafts. These will act as tires, improving traction (Figure B).

2. Glue the SPDT switches to the back of the battery holder, at the end with the wires. The switches should angle out at the 2 corners with their levers angled in toward each other, as shown in Figure C. Also, the contacts farthest from the buttons on each (the normally closed contacts) should touch. This will be the front end of our bugbot.

3. Cut the metal strip, mark enough length at each end to hold a motor, and bend each end in at about a 45° angle. This is your motor plate.

4. Examine or test your motors to determine their polarity. Tape the motors onto opposite ends of the motor plate so that their shafts point down and angle out. Orient their positive and negative contacts so that they'll spin in opposite directions.

Photography by Jérôme Demers

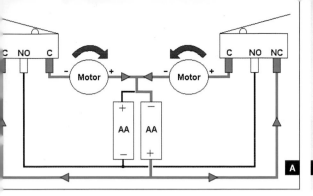

A

B

C

D

Fig. A: Diagram of Beetlebot running free, not bumped into anything; both motors draw current from the right battery only. Fig. B: Heat-shrink tubing acts as tires, giving traction to the motor shafts.

Fig. C: Switches and motor plate glued to the back of the battery holder. Fig. D: Bent paper clip threaded through a bead and glued to the rear end of the battery holder to make a rolling caster.

MATERIALS

1.5V motors (2) You can often scavenge these from toys, dollar store fans, etc.
SPDT (single pole double throw) momentary switches with metal tabs (2) You can scrounge these from an old VCR or mouse, or buy new ones for $1–$4 apiece.
Electrical wire around 22 gauge
AA batteries (2) You can also use AAAs.
AA battery holder
Spherical bead plastic or wood
Heat-shrink tubing to shrink to the widths of the motor shafts and the antennae connectors
Black electrical tape
Terminal connectors, spade type, small (2)
1"×3" piece of scrap metal plate I used aluminum.
Paper clips (4)
Cyanoacrylate (Super/Krazy) glue or epoxy
Soldering iron and solder
Toggle switch (optional) for on/off switch

FOR THE SHELL (OPTIONAL)

Round plastic lid I used a lid from a container of hair gel, but you can also use a peanut butter jar lid or anything similar.
Auto body filler putty or epoxy glue
Black and red enamel paint and primer
Clear varnish
Small, thin magnets (2) to attach shell to body

5. Use cyanoacrylate glue or epoxy to glue the motor plate down onto the back of the battery holder, just behind the switches (Figure C). Orient the motors so that the left motor spins counterclockwise as you view it from below, and the right one spins clockwise. For aesthetics, I then covered the plate with black electrical tape.

6. Unbend a paper clip, slip it through the bead, and bend it symmetrically on either side to make a caster (Figure D). Attach each end of the clip to the corners of the battery holder at the back. I used hot glue — not very professional. You could also try bending the clip ends under and soldering them to the battery connection tabs, but if you apply too much heat to the tabs, you might melt the plastic and ruin your battery holder. Beware!

Next we'll wire up the circuit, but first, an explanation: the key is that the 2 batteries work separately. Battery holders usually connect cells in series and combine their voltages, but with the Beetlebot, a wire soldered between the 2 puts them into separate subcircuits. The motors draw from only 1 battery at a time. Each switch's common connection (C) runs to a motor. The switches' normally open (NO) terminals connect together and run to the battery

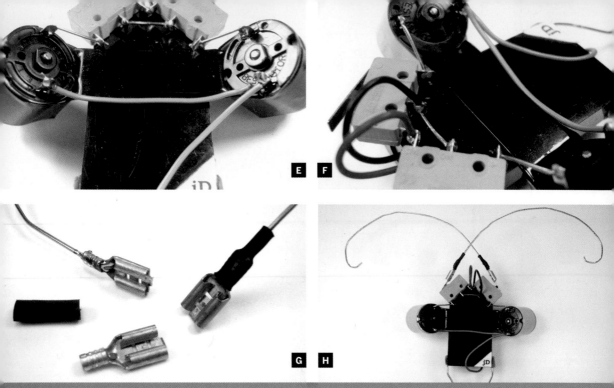

E F

G H

Fig. E: Use pieces of paper clip and insulated wire to solder connections between the switches, motors, and battery holder. Fig. F: Complete wiring, with battery holder leads soldered to switch terminals.

Fig. G: Removable antennae made from paper clips use spade connectors to slip onto switch levers. Fig. H: The bare-bones Beetlebot, finished and working, but without any switch or decorative shell.

holder's negative lead, while the switches' normally closed (NC) legs run to the positive lead.

When the bot isn't hitting anything, voltage from the positive-side battery splits and runs through both motors via the NC terminals, and the negative-side battery is not used at all. But when a switch button is activated, it closes the circuit with the negative-side battery, through the NO terminal. This reverses the motor direction on that side while the unactivated side continues running forward, which results in a quick turn away from the obstacle.

When both switches activate, both motors momentarily run backward, and the bot backs away. (The feelers cross in front, so a bump on one side activates the button on the opposite side.) That's all there is to it. Now, back to the build.

7. Solder together the 2 switches' NC terminals that are close or touching. Then solder together their NO terminals, the middle legs. I use pieces of paper clip for short joins like this, since it's faster and stronger. Then connect the common leg of each switch to the front terminal of its nearest motor (Figure E, top).

8. Solder a wire between 2 motors' rear terminals. Connect another wire from either one to any contact point on the battery holder that's electrically in between the 2 batteries (Figure E, bottom). This is the Beetlebot's all-important "third connection."

9. Finish the wiring by soldering the battery holder's positive lead to the switches' NC terminals, and its negative lead to either of the switches' NO terminals (Figure F).

10. Remove the insulation from the 2 spade connectors, and unbend 2 paper clips. Slip the connectors over the paper clips, then squeeze them down with pliers and solder in place. Dress up the connection with some wide heat-shrink tubing (Figure G). These are the Beetlebot's feelers. The spade connectors clip onto the switch levers, which makes them easy to detach for packing, and prevents damage to the fragile SPDT switches. The long paper clips give sufficient leverage to activate the switches, even if they seem hard to trigger with your finger directly.

Your robot is finished (Figure H)! Add 2 batteries, and it should come to life. If it spins in a tight circle or runs backward, you need to reverse one or both of the motor connections. To change the bot's speed or to make it run straighter, bend the metal plate to adjust the motors' angles.

Fig. I: The Beetlebot's on-off switch connects between the motors and batteries. Fig. J: Plastic lid cut to accommodate motors and antennae.

Fig. K: Building up and shaping the shell with auto body putty. Fig. L: Dime-sized masking tape circles give Beetlebot its spots.

For additional diagrams of how the circuit works, see makezine.com/12/diyscience_beetlebot.

Adding an On/Off Switch (Optional)

Every time you want to stop the robot, you need to remove the battery, which can get annoying. To solve this problem, splice a toggle switch onto the "third connection" wire between the motors and the batteries. Cut the wire, then solder in the switch and glue it to the edge of the battery holder. I neatened this connection up with more heat-shrink (Figure I).

Making the Shell (Optional)

Now here's the aesthetic part: adding the shell. I made mine out of the green plastic lid from a container of hair gel.

1. Fit the lid over the bot and cut holes in the sides to make room for the motors and the front switches/antennae (Figure J).

2. To make the shell more round, cover it with auto body putty (watch out — it cures pretty fast!) or epoxy glue, and then use files to shape and smooth it (Figure K). For final touch-up, I filled in any holes with a softer putty.

3. After sanding the lid smooth, give it a couple coats of primer, and then paint it. To make a ladybug beetle pattern, I started by painting the whole thing black (I also painted the antennae black). Then I used a dime as a template to cut round pieces of masking tape, which I applied to the lid along with a thin masking tape centerline (Figure L).

I painted glossy red over everything, and then removed the tape. For the final polish, I sanded the whole thing with very fine sandpaper and some water, which gives a glossier finish than sanding dry; this is a trick I learned from a friend who was restoring a guitar. I let everything dry and gave it 2 coats of clear varnish.

4. To connect the shell to your robot, you can glue it directly to the battery holder, or you can use magnets; glue one inside the lid and another in a matching position on the battery holder. This lets you remove the shell easily, to show your friends the insides of your biomech bug!

Jérôme Demers is a student in electronics engineering at the University of Sherbrooke in Québec. He is currently working on advanced sumo robots in both the 500g and 3kg categories.

HYDRAULIC FLIGHT SIMULATOR

Aviation trainer uses water to represent energy. By David Simpson

I volunteer with the U.S. Civil Air Patrol (the U.S. Air Force Auxiliary) teaching aviation theory and practice to junior and senior high schoolers. Two fundamental relationships I spend time on are the relationship between altitude, airspeed, and fuel, which represent forms of energy, and the one between the "four forces": lift, drag, thrust, and weight. I try to put myself in my students' shoes, and if these were explained to me in the usual way, I probably wouldn't get it.

One time I saw a desktop gadget filled with colored liquids that see-saws back and forth, and it reminded me of the exchange between airspeed and altitude, aka kinetic and potential energy. I wanted to put a control stick on that gizmo to show

the cadets I taught, and this inspired me to build a more complete flight simulator that used colored water to represent energy. So I created my Hydraulic Flight Simulator, which models the behavior of a fixed-wing aircraft in flight along the vertical or "pitch" axis.

What's neat about the HFS is that not only does it explain the relationship between fuel, altitude, and airspeed in a clear, visual way, but it also models loss of energy from drag, loss of lift from thinner air at higher altitudes, increased drag at high angles of attack, and the complete loss of lift during a stall.

You can use the tabletop simulator to demonstrate a plane's energy states through an entire flight, including take-off, climb, cruise, dive, power dive,

Photography by David Simpson and Stan Rogacki

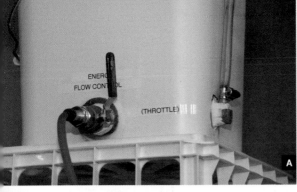

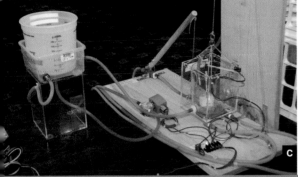

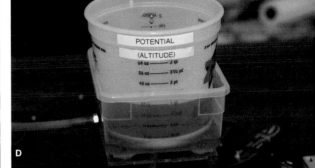

Fig. A: The energy reserve tank with flow control valve (fuel and throttle). Fig. B: Open the throttle to send energy to the Kinetic Energy reservoir (airspeed). Figs. C & D: The energy exchanger (control stick) moves the Kinetic reservoir above or below the Potential Energy reservoir (altitude), allowing energy to flow back and forth, thus exchanging airspeed for altitude in dives or climbs.

service ceiling, slow flight, power-on stall, power-off stall and recovery, descent, "dead-stick" descent, landing, and dead-stick landing. And it does all of this with all analog, easy-to-grasp components like plastic containers, surgical tubing, twine, pulleys, and some floats and valves.

Juice It Up

Here's how the system works: colored water (energy) runs down through tubing from an Energy Reserve (fuel tank) up top, into a Kinetic Energy reservoir, which represents airspeed. The rate at which it flows down is regulated by a spigot that represents the engine's throttle control (Figure A).

The Kinetic reservoir (Figure B) hangs by a pulley, and a stick control raises and lowers it above and below the level of the neighboring Potential Energy reservoir (Figure D), which represents altitude. Water/energy runs between Kinetic and Potential, depending on their relative positions.

Graduation markers on the reservoirs, made from measuring cups, serve as cockpit gauges. Increasing your relative altitude (gravitational potential energy) decreases your airspeed, and vice versa. In real life, the stick would control the airplane's pitch by adjusting its elevator.

To represent energy loss from drag, the Kinetic reservoir has little holes going all the way up its side, and the water/energy that leaks out is captured by an outer container and funneled down into a catch bucket below. The higher your speed, the more drag you get, and at some point, you just can't make the plane go any faster — just like in real life.

Similar holes in the Potential reservoir drain water slowly to account for the loss of lift at higher altitudes due to thinner air. The sim aircraft reaches its maximum altitude or "service ceiling," as in real life, when the loss in wing efficiency reduces lift to equal the downward force of gravity.

Finally, underneath it all, the catch bucket represents the ultimate thermodynamic fate of any energy expenditure: dispersed heat.

Stalls and High Angles of Attack

I also wanted to simulate stalls, which are the total loss of lift during flight and subsequent rapid loss of altitude. This is tricky — it means monitoring for a low threshold in kinetic energy and having it trigger a fall in potential energy.

To accomplish this, I used a mechanism kind of like the one in a toilet tank. A styrofoam float in the Kinetic reservoir (Figure G) connects by piano wire

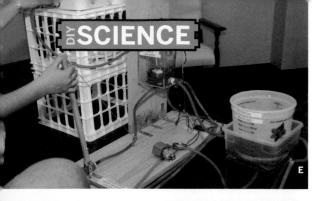

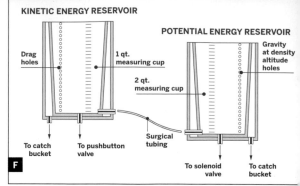

KINETIC ENERGY RESERVOIR

Drag holes

1 qt. measuring cup

POTENTIAL ENERGY RESERVOIR

Gravity at density altitude holes

2 qt. measuring cup

Surgical tubing

To catch bucket

To pushbutton valve

To solenoid valve

To catch bucket

E

F

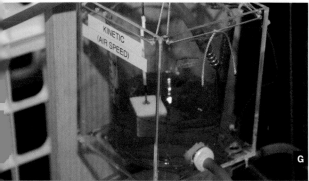

KINETIC (AIR SPEED)

G

H

Fig. E. To fly the plane, operate the throttle and the control stick together to control the flow of energy. Fig. F: Holes in each reservoir simulate loss of energy to drag (increases with airspeed) and loss of lift from thinner air (increases with altitude). Figs. G & H: To simulate stalling, a low float level in Kinetic pulls a lever that opens a solenoid valve and switches on a warning light (Figs. I & J).

up to a lever made of basswood. The lever actuates a small switch mounted to the lip of the container (Figure H). The Potential reservoir, meantime, has a drain at the bottom that runs through a solenoid valve (Figure I) and to the catch bucket.

When the Kinetic float is low, the switch routes electricity from the power supply to the valve and a "low kinetic energy" warning lamp. This simulates a stall condition; the valve opens and drains potential energy away, and the light turns on (Figure J). When the level rises, the lamp shuts off and the valve closes, allowing water/energy to be diverted to Potential. This means the airplane is going fast enough to climb.

I also added hardware to model high angles of attack, when the airplane is leaning back relative to the direction that it's going. This happens when the pilot pulls back on the stick, and it increases both lift and lift-induced drag. The pulley system already takes care of the increased lift effect, and I accounted for the greater drag by using another valve: when the stick is pulled back, a cam at its base opens a pushbutton valve that drains the Kinetic reservoir into the catch bucket. The farther the stick is pulled back, the more rapid the loss of airspeed. As the stick moves forward, the pushbutton valve closes, representing a lower angle of attack and less drag (Figure K).

Test Flight

To prepare for your flight, fill the "fuel tank" with water and add food coloring. Make sure the stick can move the Kinetic reservoir smoothly over its entire range, and turn the power supply on.

To take off, advance the throttle spigot to the fully opened position. Hold the control stick in a "neutral" position that allows the Kinetic reservoir (airspeed) to fill, elevating the float, closing the solenoid valve, and shutting off the "Low Kinetic Energy" warning light. You now have sufficient airspeed to lift off and climb.

The stick may now be pulled back (back is "nose up," forward is "nose down") to start filling the Potential reservoir, gaining altitude. Notice the effects of the drag and loss-of-lift holes. Pick an altitude that you'd like to cruise at, and as you approach it, nose down to pick up airspeed and back off a bit on the throttle, adjusting stick and throttle settings to maintain a constant altitude and airspeed. You're a pilot!

A stall condition is accomplished by allowing or causing the Kinetic level to drop below the threshold, like when you pull back too far on the stick without adding sufficient energy back in with the throttle. If you've got enough altitude, you can recover from a stall by moving the stick forward to pick up enough airspeed to regain lift.

Fig. I: At low airspeed (during stall or on ground), the solenoid valve lets energy drain from Potential. Fig. J: Warning light indicates low airspeed. Fig. K: To simulate increased drag from high angles of attack, pushbutton valve at base of stick drains Kinetic when the control is pulled back. Fig. L: To land your plane, zero your airspeed at the same moment you zero your altitude — you're back on solid ground.

Try climbing as high as you can to determine the simulator's "service ceiling." Be patient. It can be a slow process. Then find your simulator's maximum airspeed by pushing the stick forward with the throttle wide open. Unlike the real thing, your simulator won't break apart when you pass the "redline."

Periodically check the fuel gauge to be sure you have enough to complete your flight. But note that you can continue flying in glide mode, as long as you have some altitude to convert to airspeed.

To land, back off on the throttle and push the stick forward to keep airspeed safely above stall speed. The altitude reading will drop. As you get closer to the ground, ease back on the stick to slow the plane down even further, but don't get too close to the stall speed or there may not be enough altitude to recover. The goal is to stall the airplane when you're completely out of potential energy in the form of altitude — in other words, when you're right over the ground.

Future Work

My current HFS has a few obvious flaws, but what model is perfect? Here are 3 things I'd like to improve:

1. As the water level in the fuel tank drops, the pressure into the Kinetic reservoir drops as well, whereas in the real world, an engine can run just as powerfully on the first gallon in the tank as on the last. I'd like to keep the pressure consistent. One way would be to continuously pump water from the catch bucket back up to the fuel tank, while monitoring the volume pumped until the sim fuel capacity has been exhausted.

2. The size and configuration of the drag and loss-of-lift holes were not modeled for any particular aircraft — but they could be.

3. Real-world planes weigh less, and therefore perform better, as the flight progresses. That could be modeled as well.

If you have thoughts or ideas, please feel free to write me!

➕ For videos of the Hydraulic Flight Simulator in action, implementation details, materials and tools lists, and information about the U.S. Civil Air Patrol, visit hydroflightsim.net.

David Simpson (dsimpson@hydroflightsim.net) is a private pilot, and began building and flying model airplanes at age 11.

Photograph (Fig. K) by David Simpson

SHOE SHINE

Make Cinderella jealous. By Norene Leddy and Ed Bringas, with Andrew Milmoe and Melissa Gira

Platforms, the latest series of work in the ongoing Aphrodite Project (theaphroditeproject.tv), is a social sculpture: an interactive, wearable device that is a conceptual homage to the cult of the Greek goddess, as well as a practical object and a vehicle for public dialogue.

An integrated system of shoes and online services, *Platforms* draws on the innovations of the courtesans of antiquity to improve the conditions of 21st-century women who, despite advances in culture and technology, are still vulnerable to surveillance and violence. *Platforms* empowers all women with tools they can make to stay safe.

While we designed the shoes with women's safety in mind, they're also intended as a vehicle for self-expression. The video shoe is essentially a portable media player (PMP) placed in a platform shoe, and

can be configured to show whatever movies, images, or messages you put on it. The other shoe has a 120dB siren, for when you feel like sounding the alarm.

1. Make balsa templates for the electronics.

1a. Make a balsa wood block the size of your PMP plus ⅛" on the face and ¹⁄₁₆" on all other sides.
1b. Unscrew the alarm casing and remove the piezo speaker and circuit. Make a balsa block the size of the circuit and speaker with the 9V battery and snap connector, plus ¹⁄₁₆" of wiggle room on all sides.

2. Prepare the shoes.

2a. Trace the outlines of the balsa wood blocks on the outside of the heels, centering so at least ¼" of the heel is intact on all sides (Figure A). Using an

Photography by Norene Leddy and Ed Bringas

Fig. A: Balsa template for alarm components, wrapped in foil and traced onto heel. Glue this foil so it won't slip. Fig. B: Template and hollowed heel wrapped in foil. Cover exposed parts of the shoe; ShapeLock is tough to remove. Fig. C: Template aligned with foot bed and heel. Once this angle is set, fill the rest with ShapeLock. Fig. D: Check the angle again, making sure shoes match. Smooth and fill gaps with ShapeLock and a spoon.

MATERIALS

Platform shoes with 4" or higher heel
Leather to line cork heels
Piping (purple trim on page 148)
Super glue and rubber cement
22 gauge (AWG) wire
Balsa wood
Shoe glue
12V DC piezo siren audible alarm
  RadioShack part #273-079
9V battery
9V snap connector
Clear ⅛" plexiglass
¼"×6" very thin sheet steel **must be easily cut**
Mylar, clear or frosted
ShapeLock thermoplastic, 500g shapelock.com
Small portable media player (PMP)
  We recommend the iRiver CXW-2G Clix.
¼" diameter rare earth magnets (6)
SPST "Soft-Feel" push on-push off switch
  RadioShack part #275-1565
Heat-shrink tubing RadioShack part #278-1611
Hair dryer and lighter
Leather punch
Metal or nonstick saucepan
Meat thermometer
Metal tongs
Dremel tool
Olfa knife
Plexiglass knife

Olfa knife, hollow out the heels around your outline plus an additional ⅛" all around, or more if possible, to make room for the ShapeLock plastic housing. The more you can cut out for the ShapeLock, the sturdier your shoes will be when you wear them.

2b. Line the hollow spaces of the heels with aluminum foil as a release for the ShapeLock. Wrap the balsa blocks in foil (Figure B). Glue the foil so it won't come off the blocks.

3. Prepare the ShapeLock.

Heat water to 160°F in a saucepan. Drop about a quarter of the ShapeLock into the water. When the ShapeLock changes from opaque to translucent, remove with tongs, and let dry.

Give yourself 30 seconds or more before handling so you don't burn yourself, then knead out any excess water. ShapeLock has a working temperature of 150°F, which should give you 2–5 minutes of working time. Try a few tests to get a feel for the material.

4. Apply the ShapeLock and place the wooden templates.

The hardest part of making these shoes is aligning the PMP with the outer edge of the foot bed and heel. Most platform shoes taper inward from the foot bed,

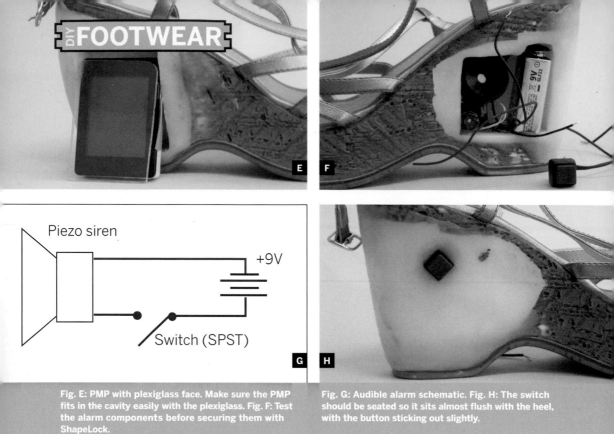

Fig. E: PMP with plexiglass face. Make sure the PMP fits in the cavity easily with the plexiglass. Fig. F: Test the alarm components before securing them with ShapeLock.

Fig. G: Audible alarm schematic. Fig. H: The switch should be seated so it sits almost flush with the heel, with the button sticking out slightly.

so your PMP will need to rest at an angle.

4a. Line the hollowed-out heel with ShapeLock, allowing enough space to insert the PMP template block. While the ShapeLock is still soft, place the wood block flush with the foot bed and heel (Figure C). It may take a few tries to get the right angle. If it doesn't work, remelt the ShapeLock and try again.

4b. Allow the ShapeLock to harden completely, and then remove it and the aluminum foil from the shoe. Hardened ShapeLock will not bond with any fresh ShapeLock or with the shoe, so melt the outer layer with a hair dryer and then place it back in the heel to bond with the shoe. Tiny gaps can be filled in later. With the hair dryer, melt the outer layer of ShapeLock and fill in the rest of the heel, heating and cooling as needed.

4c. Repeat with the alarm shoe. The placement of the alarm template is not as critical as the placement of the PMP, since leather will cover the entire speaker. Just make sure that you leave enough room for the ShapeLock to have a minimum of ⅛" thickness, and that it's not sticking out from the heel or foot bed.

4d. Double-check your heel measurements, making sure that both are even, and that the heel dimensions have not changed. Melt more ShapeLock as needed

to fill in any gaps, and heat the hardened ShapeLock before applying new ShapeLock. For a final shaping, place a metal spoon in the saucepan to heat up. Use the hair dryer to heat the ShapeLock, then use the heated spoon to smooth it (Figure D). With a screwdriver or knife, remove the balsa blocks from the heels.

5. Install the PMP.

Cut a piece of ⅛" plexiglass to cover the face of your PMP (Figure E). Make sure the player fits in the ShapeLock heel with the ⅛" plexi. This may require some sanding. If needed, use a Dremel at low speeds to remove excess ShapeLock. If it gets too soft during sanding, put the shoe in the freezer to harden it between sandings.

6. Install the audible alarm system.

6a. Dip the alarm circuit in rubber cement or rubberized glue to protect the circuit. Once dry, arrange the speaker, circuit, and battery in the heel (Figure F), and mark an area to drill a hole for the switch on the opposite side. Make sure the drilled hole won't interfere with the circuit board or speaker.

6b. Drill a hole through the ShapeLock large enough for the diameter of the switch. Using a Dremel tool, mill out enough room for the top of the switch to

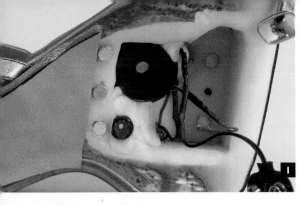

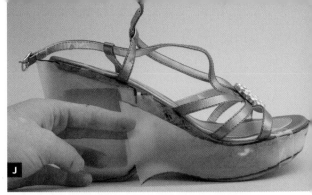

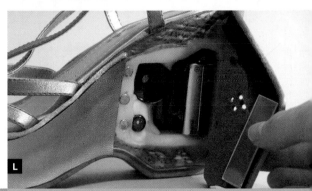

Fig. I: Secure the speaker and circuit board with ShapeLock, but leave the battery and snap connector loose. Fig. J: Wrap the mylar around the heel and trace the outline with a Sharpie to create the pattern for your leather. Fig. K: Mylar pattern for PMP flap includes a cutout for the PMP screen. Fig. L: Completed alarm system with holes punched in leather flap and sheet steel glued to the edge for magnetic closure.

sit almost flush with the heel. Connect the switch to the circuit and battery to test (Figure G).

6c. Connect the wires, and slip heat-shrink tubing over the exposed connections. Using a small lighter, place the flame near the heat-shrink tubing to tighten the tubing to the wires.

6d. Use a small amount of ShapeLock to secure the button, circuit, and speaker in place (Figures H and I). Remember to heat the ShapeLock inside the shoe so it will bond. Then drill 2 holes the size of your magnets behind where the battery will sit. Super glue the magnets in place; these will hold the battery.

7. Finish the exterior.

7a. For each shoe, trace a mylar pattern for the leather to wrap about ¾ of the sole, starting at the center of the heel, along the inside of the shoe, around the toe, and ending with a seam ½" to the front of the PMP or alarm cavity (Figure J).

7b. Use this pattern to cut the leather (for an example, see our pattern at makezine.com/12/diy_footwear); for the alarm shoe, also cut a small hole for the switch. Tack the piping to the top edge of the leather with rubber cement, if the piping bunches around the curves, notch its backside. Attach the leather with shoe glue, from the front side of the electronics cavity all the way around to the center of the heel.

7c. Now make a second pattern for each shoe, going from the center of the heel and covering the electronics, overlapping the first pattern ½" to the front (Figure K). This is the flap that will allow you to remove the PMP and the alarm battery.

Cut out the PMP flap, then cut out a window for the PMP screen, leaving space to attach the plexiglass to the leather. Attach piping to the top and seam the edges, then glue the flap along the center of the heel, allowing the leather to hinge. Glue the plexiglass cover to the leather, making sure it sits flush when closed.

Cut out the alarm flap, attach piping, and use a knife or punch to make small holes over the speaker to let sound out. Glue along the center of the heel.

7d. To close the flaps, glue a thin steel strip to the front edge inside each flap. Drill 2 small holes the size of the magnets into the ShapeLock, aligned with the steel strip. Super-glue the magnets in place. These will hold the flaps closed, and allow you to remove the PMP or battery (Figure L).

The Aphrodite Project is a series of multimedia artworks — *Sanctuary*, *Platforms*, and *Kestos Imas* — started by Norene Leddy in 2000. theaphroditeproject.tv

THE WIDOWMAKER: CUTTING DOWN A TREE

By Tim Anderson

❖ Screw your courage to the sticking place ... ❖
—William Shakespeare

⚠ **WARNING! Get really drunk first. Then it won't hurt so much when you chainsaw your face off and crush your family.**

Step 0. Find a tree to cut down.

No problem. As soon as you know how to cut down a tree, people will sense it and constantly ask you to cut down trees for them.

Here's some helpful technical vocabulary:

» **Equipment** Every piece of equipment used in this process is called a "widowmaker."

» **Tree anatomy** The tree, and every part of the tree, is called a "widowmaker."

» **Terrain analysis** Everything in the vicinity of the tree is called a "widowmaker."

Act really confident and relaxed, as shown in Figure 0. I'm saying, "You mean this tree over here?"

Step 1. Cut a notch in the felling direction.

Make it a big notch that goes more than halfway through the tree. The "mouth" part of this notch faces the direction you want your tree to fall.

If the tree is mostly upright and its branches are sort of symmetrical, it will want to fall in the direction of the notch mouth. Tie a rope to the top of the tree to help it fall in the direction you want.

Step 2. Cut toward the notch from the other side.

Cut from the other side toward the big notch. It's just a single cut straight toward the big notch, leaving a thin "living hinge" to make an axis of rotation to control the trunk's fall. When you get close to the notch, the tree will start to lean away from you. If it doesn't, or if it leans the wrong way and binds your saw blade, have the Oompa-Loompas pull on the rope tied to the top of the tree.

Step 3. The tree falls.

As the tree starts to lean, make your escape to one side, because the butt of the tree can kick backward. There isn't any very safe place to be, because huge branches can break and fly anywhere, or the top of the tree could pull something down with it.

Photography by Moana Minton

ODE TO THE SWEDE SAW

In the lower left of Figure 4, you see a couple of hand saws on the ground. Those are old bow saws, also called "Swede saws" because of where they were invented.

One of our old Minnesota neighbors cut firewood for a living with a big old crosscut (cuts on both strokes) two handed handsaw in the 1920s. He was just getting by, selling 4 cords of wood a day. Then he bought one of the new Swede saws, started cutting 8 cords a day, and had enough money to get married.

The tubular steel bow puts the blade under high tension so the blade can be very thin without puckering. Because the blade is narrow, it won't bind in the kerf (groove made by a cutting tool) as much. Bow saws can cut a very narrow kerf, removing less wood, and do it fast with less work than previous saws, which weren't much different from what the Romans had.

My cousin's father-in-law was killed by a vine that was pulled by a tree as it fell.

If you didn't cut all the way to the notch, your "living hinge" may keep the tree from falling to the side.

Step 4. The tree's hung up. Now what?
I misjudged the height of the tree and the distance to its nearest neighbor. I have no depth perception. That's why they wouldn't let me fly jets. It fell in the right direction, but it skinned the next tree over and hung up in it. Now it needs to be sawn through the middle to finish falling down.

There are always problems like this. This one's called a "widowmaker" because you've just compressed the spring of a giant trap, and now you've got to walk into it and saw through the trigger.

Step 5. Saw up from the bottom.
There's no safe way to proceed once you've got a "leaner" tree hanging like a compressed spring. If you cut down from the top, the tree will sag and bind the blade like a clamp. This is dangerous, because if the blade catches, then the chainsaw kicks back at you and can alter your appearance. Basically, you need a safe way (there isn't one) to cut the middle of something supported on both ends.

Here's my cousin Rod's method: Saw up from the bottom until the tree starts to sag. Then saw down from the top until a good outcome ensues. Cut a notch if the blade starts to bind.

Step 6. Check yourself into the clinic.
I stand in awe at the sight of this forest giant laid low. If things don't go so well, your loved ones will get to contemplate your own mortality.

Tim Anderson (mit.edu/robot) is the founder of Z Corp. See a hundred more of his projects at instructables.com.

SPECIAL THANKS TO *VICKY* AND *GARRETT!*

LAID OUT LIKE SO, ALL *I* HAD *TO DO* WAS *EXECUTE.*

BEING WELL *PREPARED,* I WAS *CONFIDENT.* SUCCESS WAS A MATTER OF PATIENCE AND FOCUS.

I MADE THE CUTS *CAREFULLY,* THE FOLDS *PRECISELY.*

WITH CONFIDENCE I COULD NOW *IMPROVISE.* I MADE SLEEVES AND *TAPED* THE SEAMS SHUT WITH *DUCT TAPE.*

THE COAT WAS A PERFECT FIT.

NOW, WALKING THE STREETS I *CONTEMPLATED* THE *POSSIBILITIES.*

ALL THE DIFFERENT COATS THAT WERE *WITHIN* MY REACH.

TRASH BAG RAIN-COAT

WHERE *OTHERS* HAD SEEN ONLY TRASH BAGS,

DUCT-TAPE SEAMS

RECEPTACLES *FIT* ONLY TO RECEIVE OUR *WASTE,*

I HAD SEEN SOMETHING DIFFERENT...

AND *IT* WAS BEAUTIFUL IN ITS SIMPLE *FUNCTION.*

CUT SLITS AND STRIPS FOR BELT

THE *RAIN* LIGHTLY MASSAGED MY BRAIN.

IT WOULD BE *IMPERVIOUS* TO THE ELEMENTS THAT WOULD *IMPRISON* ME.

I WAS FREE.

DRAGOTTA GRIFFITH, BONSEN

APOLOGIES TO *FRANK MILLER!*

MakeShift

By Lee D. Zlotoff

The Scenario: You are alone for the weekend in your vacation home, where a 40-foot-tall, white-fly-infested tree partially blocks your otherwise glorious view. To save the cost of calling in a professional, you've decided to finally get rid of it yourself. You won't have to worry about the noise bothering anyone, since your nearest neighbors are at least a half-mile away. From some internet research, you've learned that you'll need to cut off most of the branches first, then notch the base properly and set a wedge to make sure it falls in a safe area and doesn't end up on your house. Your electric chainsaw should take care of the larger lower branches and, once they're cleared, you can do the smaller and higher ones with your pole saw. So you get out your 20-foot extension ladder, set it securely against the trunk, and, looping the saw's extension cord over your shoulder to keep it clear of the blade, climb up, chainsaw in hand.

Your work starts smoothly, with some of the larger, lower branches breaking off under their own weight after you cut partially through them. The fallen branches pile up on the ground below. As you go higher and the branches get lighter, you have to cut deeper into them. The saw is beginning to feel heavier, and it's time for a water break.

But as you shut down the saw and start back down the ladder, you lose your balance. You instinctively drop the saw to grab for the ladder with both hands, but it's too late. You end up taking a nasty 15-foot fall into the tangled pile of branches below, and — damn it all — right onto a 4-foot-long, 1-inch-wide stob, which, like a giant punji stick, *passes straight through your thigh next to the bone!*

The pile of branches helped break your fall. But the branch you've been impaled on is locked into the other branches like a bunch of springy pick-up sticks, and together they weigh a great deal. And it doesn't help that the bark is rough, and a few smaller branches attached to "Vlad" also have passed through your leg and now act as barbs, making extraction very difficult at best. There's not much blood visible, but you know you need to do something for yourself fast.

What You Have (and Don't Have): You're fully dressed as any good woodsman would be. Given the distance of your neighbors, shouting would be a waste of energy. Your cellphone and the land line are both in the house, as are your Swiss Army knife, first aid kit, and the keys to your car, parked out front in the gravel driveway. The chain bar is now bent and the chain has slipped off in the fall, so even though you can reach your chainsaw, it's a goner. The pole saw is 10 feet away, resting against a chair, with the handle in the upward position pointing away from you, and with its cutting blade resting on the ground, the way anyone interested in safety would place it. Meanwhile, this sucker in your leg is really starting to hurt ... so now what?

Send a detailed description of your MakeShift solution with sketches and/or photos to makeshift@makezine.com by Feb. 29, 2008. If duplicate solutions are submitted, the winner will be determined by the quality of the explanation and presentation. The most plausible and most creative solutions will each win a MAKE sweatshirt. Think positive and include your shirt size and contact information with your solution. Good luck! For readers' solutions to previous MakeShift challenges, visit makezine.com/makeshift.

Lee D. Zlotoff is a writer/producer/director among whose numerous credits is creator of *MacGyver*. He is also president of Custom Image Concepts (customimageconcepts.com).

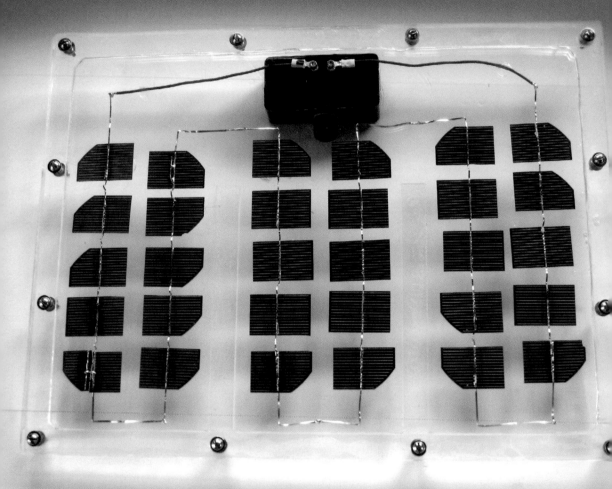

20-Watt Solar Panel

Using a few solar cells and a plastic case, you can utilize the sun's energy to power anything from a light bulb to your entire house. By Parker Jardine

The sun is a renewable energy source that's free and plentiful. Some people power their entire home with solar energy. A few even sell back the energy to the electricity grid for a profit.

I decided to start small and build my own solar panels to supplement my workshop power needs. Here, I'll explain in detail how to build a 16.5-volt, 20-watt solar panel.

In the next volume of MAKE, I'll show you how to integrate the solar panel(s) into your electrical system.

>>

MATERIALS

[A] ½" watertight compression wire connector

[B] Copper 18AWG hookup wire

[C] Ring tongue terminals (2)

[D] 20-watt DIY Solar Panel Kit **part #DIY20W, about $49. Available at siliconsolar.com.**

[E] ¼" stainless steel split lock washers (24)

[F] 1½"×¼" stainless steel pan head bolts (12)

[G] ¼" stainless steel flat washers (24)

[H] ¼"-20 stainless steel finished hex nuts (12)

[I] ¼" acrylic cut to the following sizes: 2 sheets 18"×24", 2 pieces 1"×16", 2 pieces 1"×24", 2 pieces 1"×10" **You can buy custom-cut acrylic at a local glass company, or you can order McMaster-Carr part #8589K83, which you'll have to cut to size.**

[J] Ribbon wire **for connecting solar cells, Silicon Solar part #04-1010**

[K] 60/40 rosin-core solder, .05" diameter **RadioShack part #64-006**

[L] Devcon Duco Cement **McMaster-Carr part #7447A16, or similar clear plastic cement**

[M] GE Silicone II sealant **or similar clear silicone sealant**

[N] RTV silicone adhesive **found at auto parts stores**

[O] Rosin soldering flux 2oz non-spill paste **RadioShack part #64-022**

[P] Chassis-mount dual female binding post **RadioShack part #274-718**

[Q] Project enclosure 5"×2½"×2" **RadioShack part #270-1803**

TOOLS

[R] Soldering iron or gun **Gun is preferred.**

[S] Voltage/amp meter

[T] Soldering helping hands

[U] Wire cutters

[NOT SHOWN]

Power drill with ¼" and ¾" bits

Jigsaw (optional) **only if cutting your own acrylic**

Acrylic cutting knife (optional) **also used when soldering the cells**

Flathead screwdriver

One-handed bar clamps

1. BUILD THE PLASTIC CASE

1a. To start, cut the acrylic to the following sizes: 2 sheets 18"×24", 2 pieces 1"×16", 2 pieces 1"×24", 2 pieces 1"×10".

If you purchased the acrylic from McMaster-Carr, it comes in 24"×24" sheets; you'll need 2 of them to make all the parts. You can cut them with a jigsaw, but I recommend that you have your local glass company do all your custom cutting. Acrylic is difficult to cut because you have to use the right blade, and cut at the right speed. Many people end up cutting too fast; this results in the acrylic melting and then binding back to itself after the blade has passed.

An alternative is to cut the acrylic with a special acrylic cutting hand blade, which is good for acrylic 1/8" thick or less, but very difficult on our 1/4" thick sheets. The method is to score the acrylic on the desired cut line multiple times, then place the sheet on the edge of a solid surface and press down to snap the section off from the rest of the sheet. It's easier if the score line is farther away from the edge, so that you have more room to press down to snap the sheet.

1b. Peel the protective cover off one side of one 18"×24" sheet. On this exposed side, fit together the two 1"×24" pieces and the two 1"×16" acrylic pieces to build a border, and then use the Duco cement to bond the border onto the sheet. Space the two 1"×10" acrylic pieces evenly across the case, along the bottom 24" border; I spaced them 6⅝" apart from each other. Note their position, and then use the Duco cement to bond them to the sheet. Let the cement dry according to the directions on the tube.

1c. The dual binding post comes with a mounting template on the back of the package. Cut out this template and use it to mark the 2 drill holes on the acrylic sheet, ¾" apart and ½" from the top border edge (see Figure B for placement). Then use an 11/64" or 3/16" drill bit to drill both holes through the acrylic sheet. Insert the binding posts from the back of the acrylic sheet, then tighten the nuts on the binding posts on the inside of the case. You'll have to clip off the tops of the binding posts to close the case with the top acrylic sheet.

1d. Place the second 18"×24" acrylic sheet on top of the case, lined up with the edges of the other acrylic pieces. Tape this top sheet down to keep it from moving, and then clamp the whole case with

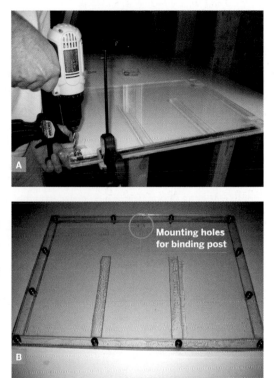

Mounting holes for binding post

at least 2 one-handed bar clamps to keep it steady during the drilling process.

1e. Using a ¼" drill bit, drill holes through the case border in 12 evenly spaced locations (Figures A and B). After each hole is drilled, add its stainless hardware. This ensures that the case will line up evenly when you're done drilling it. You don't want the top sheet to shift at all during this process.

I also recommend that you start in one corner, and next drill the opposite corner. After hardware has been placed in each corner, I then drill 2 more holes along each side. This leaves a total of 12 bolts in the case. I do this to guarantee that the case won't leak between the top sheet and the middle border. Congratulations, you now have a waterproof plastic case!

2. SOLDER THE SOLAR CELLS

2a. If you haven't soldered solar cells before, I recommend you try soldering cells that are considered scrap. You can buy scrap solar cells, by the ounce or by the watt, at siliconsolar.com. To brush up on your soldering skills, see the Make Video Podcast soldering tutorial at makezine.com/go/solder.

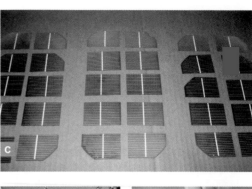

Cell Alignment and Tab Ribbon Length

Line up the 30 solar cells in 6 columns of 5 cells each on top of cardboard or fiberboard (Figure C). Arrange the cells in your desired layout. Measure the distance from the top edge of one cell to the bottom edge of the cell directly *below* it in the same column. You'll cut the ribbon wire this distance plus ½". After cutting the tab ribbons, bend each ribbon in the center to provide flexibility between each cell.

Tinning the Tab Ribbons

Tinning is a process where you apply solder to the length of the tab ribbon. This is a crucial step in the soldering process. Simply take the tab ribbon, and clamp one end in the helping hands tool. Heat up your solder gun and apply solder onto the topside of the tab ribbon. Apply only a small amount of solder, and try not to glob it on. Make sure you tin the appropriate edge of all tab ribbons before trying to solder the ribbon to the cell (Figure D).

Preparing the Solar Cells

Apply a small amount of flux to the surface of the solar cell where you'll solder the tab ribbon (Figures E and F). Make sure to wipe off excess flux, because too much flux can prevent your tab ribbon/solder from bonding to the solar cell. It will also decrease the efficiency of the cell if a layer of dried flux is blocking the sun. Some soldering experts say it's a good idea to scuff up the area where you plan to solder; this scuffed-up surface may make it easier for the tab ribbon and solder to bond to the cell.

Connecting the Rows

When assembling each row of 5 cells, make sure you solder the bottom or top cell with a longer piece of tab ribbon. This ensures that the row can connect to its adjacent row without adding more than one solder joint.

NOTE: Wait until all rows are soldered and placed permanently in the case with the RTV silicone before soldering the connections between rows.

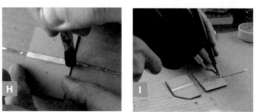

2b. Start by soldering the tinned edge of a tab ribbon to the topside (shiny side) of the first solar cell. Lay a tab ribbon halfway up the cell, with most of its length extending up beyond the edge of the cell. Make sure to solder the entire length of the tab ribbon. Now solder the second, third, fourth, and fifth solar cells exactly like the first one (Figure G).

2c. Next, flip over the 5 solar cells. The tab ribbon that extends above one cell will now be soldered to the backside of another cell. Solder all 5 solar cells together connecting each topside to the backside of the next cell (Figures H and I).

Verify that you soldered the cells together correctly, by using the voltmeter to test the positive and negative ends of the newly soldered row of solar cells. Face the topside of the cells into the light and test the voltage. The voltage should be approximately 2.5V (the actual voltage will be slightly higher). Be sure to test the voltage of each row of soldered cells (Figure J).

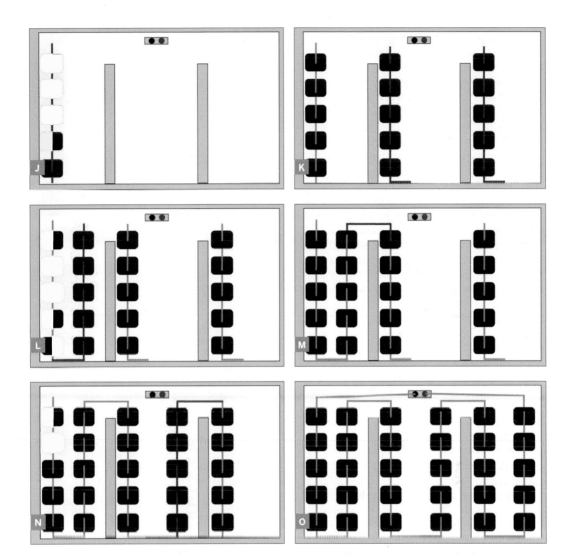

2d. Now that you've connected 5 cells to form 1 row (Figure J), copy this exact process for 2 more rows, using a total of 10 more cells. Place the 3 newly soldered rows on the left side of each section inside the case (Figure K).

2e. Solder 3 more rows in the same manner as the first 3; these will be the right-hand rows. Now rotate them 180 degrees, so their voltage runs opposite to the left-hand rows. You'll solder the ribbon wire from each left-hand bottom cell (positive voltage) onto the top (negative) side of each right-hand bottom cell (Figure L), and you'll turn the corner at the tops of the rows in a similar way (Figure M) — but don't

solder yet. First, place all 3 right-hand rows in the plastic case and verify their position in relation to the other cells (Figures N and O).

3. INSTALL THE CELLS AND WIRE THE CASE

3a. After all the rows have been set into the case, apply RTV silicone adhesive to the back of each cell. I usually apply a generous amount in 2 or 3 spots (Figure P). Be careful not to crack or break the cells. After applying the silicone, place the row into the case at the desired location. After the silicone has dried, the cell cannot be removed from the case.

3b. After all the rows have been glued into the case, solder the connection between each row. Use a piece of cardboard to keep from damaging the acrylic with the solder tip (Figure Q).

3c. Cut 2 hookup wires to length, extending from the end of the positive and negative tab wires to the binding posts. Crimp a ring tongue terminal to each wire, solder the opposite end to the tab ribbon, and then tighten the terminals onto the binding posts. Apply some RTV silicone to the hookup wire, so it won't move around in the case. You can also just use more tab ribbon instead of the hookup wire to connect the solar cells to the binding posts (Figure O).

Before sealing the case, test both the positive and negative binding posts in the sun to verify that you have the appropriate voltage. The voltage should be approximately 16.5V.

4. SEAL THE CASE

Apply a generous amount of silicone sealant all the way around the border of the case. Make sure you apply it around the holes where the stainless steel hardware will hold the case closed (Figure R).

Place the top acrylic piece on the top of the border to seal in the cells. Now fasten the stainless steel hardware to the case.

5. INSTALL THE JUNCTION BOX

5a. First, cut off the closed end of the project enclosure case. Make sure you cut it straight. You can use a jigsaw, table saw, miter saw, or handsaw. Drill a ¾" hole into the side of the project case (Figure S).

5b. The ¾" hole in the junction box is for the ½" watertight compression wire connector. This connector can be used with 14/2 or 12/2 outdoor-rated wire, or flexible conduit as seen in Figure X.

Many watertight connection options are available, including metal conduit.

This solar panel does not include a diode to protect the panel from reverse current. If you're not using a charge controller between your batteries, you'll need to add a blocking diode to your solar panel. This prevents reverse current flow back to the panel at night.

5c. Place the project box over the exposed binding posts on the back of the solar panel. Note your desired position, apply a generous amount of Duco

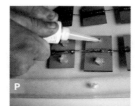
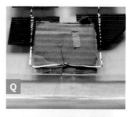

cement to the cut edge of the project box, and then place the box in position. Let the cement dry for a few hours, then apply a generous amount of silicone sealant around the edge of the box, inside

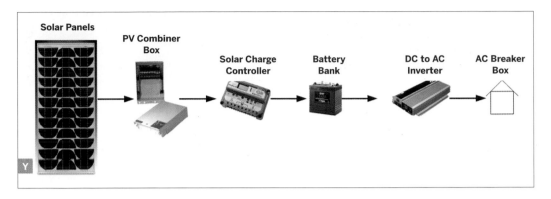

| Solar Panels | PV Combiner Box | Solar Charge Controller | Battery Bank | DC to AC Inverter | AC Breaker Box |

and outside, to ensure a watertight seal (Figures T, U, and V). Your charger is built (Figure W). Now it's time to put it to work.

MOUNTING AND GROUNDING

Mounting Options

I recommend mounting solar panels onto 2 aluminum rails. Place the aluminum rail against the stainless steel bolts, hammer the rail, then flip the rail over; you'll now see the drill locations needed.

I've also used UniRac's SolarMount modular mounting system; I purchased 2 rails and 2 tilted legs to create a ground mount system that can also be used on a flat roof. Drill holes in the rails to match the mount on the UniRac mount system (Figure X).

Orientation

It's recommended to orient solar panels toward due south. Simple enough. You also want to angle your solar panel perpendicular to the sun. However, the sun follows the azimuth angle throughout the day — the greatest charging potential is during solar noon — and the sun's angle changes not only throughout the day, but also throughout the year.

If you're going to mount the solar panel in a fixed position, a good rule of thumb is as follows. If you are off-grid, mount the solar panel for latitude plus 15 degrees. Winter is the season that's most critical for off-grid homes, because of the lack of sunlight.

If you have a grid-tie setup, mount the solar panels for latitude minus 15 degrees. This orientation will increase your generation during summer months, when you can sell more power back to the grid, or use it for air conditioning.

If your solar mounting system is adjustable, adjust the panels at least 4 times a year, bringing them

(roughly) perpendicular to the sun angle at solar noon. Go to susdesign.com/solpath to see tools related to the solar path for your particular latitude.

Grounding

According to the National Electrical Code, all solar panel systems need to be grounded to code. I recommend reading the NEC handbook for detailed information regarding electrical codes. At a minimum, the NEC requires that all systems must have equipment-grounding conductors that connect the metal surfaces of the solar panels to a ground rod. My grounding system consists of the following items:

» Two code-compliant ground lugs screwed into each aluminum rail that touches the solar panel.
» An 8-foot grounding rod driven into the earth near the solar panel array.
» 6-gauge stranded copper wire (preferred) or solid wire connected to the ground lug and then to the ground rod.

Also read John Wiles' article, "To Ground or Not To Ground: That Is Not the Question (in the USA)," available at makezine.com/go/wiles.

System Design

In the next volume of MAKE, I'll show you how to connect the solar panel to your electrical system. For now, check out Figure Y for a visual overview of the electron flow in a typical solar system.

Parker Jardine is a systems administrator for Durango School District 9R in Durango, Colo. He enjoys cycling, kayaking, rock climbing, electronics, and renewable energy.

TOOLBOX

Holiday Kits for Makers

Makers are not typical when it comes to the holidays. They don't want to be buried in packaging; they'd rather give or receive something they can build themselves. And they like stuff they can personalize to make their own. Most of all, makers like *useful* presents, whether it's something they'll use on a daily basis or a tool that makes building that totally impractical project a little more … practical.

Here's a roundup of kits, books, web services, and tools we think the makers in your life will like. Be sure to check out our online gift guide at store.makezine.com for more great finds as the holiday season gets into full swing.

Hydrogen Fuel Cell Kit
$125 makezine.com/go/fuelcell

When I was a kid, I lived for all the science and tech kits I'd get for Christmas: the chemistry sets, the rocketry starter kit, the telescope set, the X-Acto hobby tool chest. Today's budding nerds get to experiment with hydrogen fuel cells, thanks to this great kit. Experiments include: How to build a solar-powered car; Effects of direct and indirect radiation; Electrolysis and its effect on water; Oxy-hydrogen test; How to construct and load a reversible fuel cell; Decomposition of water in the fuel cell; Qualitative and quantitative analysis of gas in a fuel cell; How efficient is electrolysis?; How light influences electrolysis; Solar electrolysis; Fuel cell-powered car. I wonder if it's too late to finagle one of these under my Christmas tree?
—*Gareth Branwyn*

Tuna Tin Kits

$25 qrpme.com

There's nothing fishy about these QRP amateur radio kits housed in Two Tinned Tunas cans. You can buy one with all the bells and whistles, or get the bare-bones kits, which require some additional parts scrounging. Or check out the Picaxe Construction Set, a little self-contained Picaxe microprocessor breadboard system.

Mugi Evo Plane Kits

From £23/$47 mugi.co.uk

These Mugi Evo twin-wall polypropylene glider kits look fun, and they're pretty inexpensive. They're made from corrugated plastic, the stuff used for lightweight signs and produce boxes. Mugi also offers cool LED lighting options, and it's easy enough to add lightweight servos.

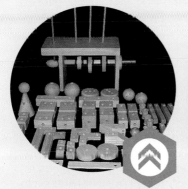

Awesome Automata

$60 store.makezine.com

Available in our very own Maker Store, the Designing Automata Kit teaches about simple mechanics using cams and a crank slider mechanism. (No glue needed!)

Many different designs can be made, and the kit can be used over and over again. MAKE is proud to be the only store this side of the pond to carry this kit, produced in Thailand using chemical-free rubberwood from sustainable sources.

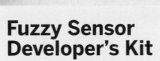

Fuzzy Sensor Developer's Kit

$70 ifmachines.com/products.html

This kit offers an up-close look at IFM's fuzzy sensor technology, which makes the fabric itself a sensor by using electronic yarns and materials. Designed specifically for toy, fashion, and other electronic product developers, it's a cool way to play with soft circuits. MAKE readers can get an additional 10% off through the holidays by entering passcode "make" at the checkout!

Classic Radio Kits
Prices vary

We can't forget to include a couple of radio kits, like a World War II foxhole radio kit (which uses a razor blade instead of a crystal) from xtalman.com. Or check out the shortwave kits at radio.tentec.com/kits, and Niel Wiegand's great how-to at makezine.com/go/tentec1054.

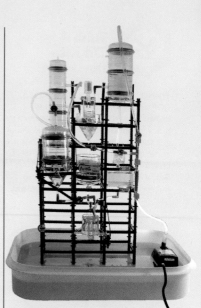

Hydrodynamics Kit
$90 makezine.com/go/hydro

This 300-piece hydrodynamics building set from ThinkGeek includes girders, tank parts, tubing, valves, tube connectors, meters, and an electric pump with AC adapter. We like it because it's not a cut-and-dried kit; you have to actually think the building process through!

Gyroplane Flying Motorcycle Kit
$39,995 makezine.com/go/skycycle

The Super Sky Cycle kit ain't cheap, but it has a 300-mile range and a top speed of 100mph flight/55mph highway, and can carry a pilot up to 280lbs. It's freaking awesome, and "street legal," whatever that means here.

Boarduino
$18 makezine.com/go/boarduino

Wow! Check out the Boarduino, a breadboard-compatible Arduino clone (or see the Maker Store for our own Arduino kit). You'll never have to struggle with a solderless breadboard again. When programmed with the Arduino bootloader, it can talk to the Arduino software and run sketches just like the original.

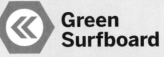

Green Surfboard

$200 greenlightsurfsupply.com

The handmade surfboard is a classic DIY project. Here's an eco-friendly kit from Greenlight: a recyclable EPS foam core, bamboo fins and stringer, bamboo fabric instead of fiberglass, bio-plastic leash plug, and low-VOC epoxy. It comes with a set of DVDs to show you how to build it.

 # ROV in a Box

$250 nventivity.com/roviab.html

Save yourself a million bucks or so and try out this underwater robot kit. Includes pretty much everything you need for a working robot, including the frame and lights for night diving.

Tricks of the Trade By Tim Lillis

One small lock and two bikes? No problem!

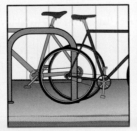

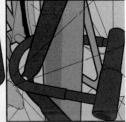

Stuck with two bikes and one lock? And it's a mini-lock? Fear not, this trick from Ethan Clarke at Refried Cycles in San Francisco will be your savior.

First, arrange the bikes at a suitable lock post facing opposite directions with the rear axles as close as possible to the post.

Next, near the dropouts, where the seat stay and chain stay meet, thread the lock through. Position the curved end of the shackle within the plane of one of the wheels, and pivot the crossbar down to meet the other end.

Rest easy!

NOTE: As is usually the case with a mini U-lock, your wheels are vulnerable, so get some locking skewers.

Have a trick of the trade? Send it to tricks@makezine.com.

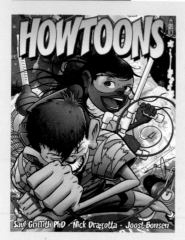

« Kids Rock

Howtoons by Saul Griffith, Nick Dragotta, and Joost Bonsen
HarperCollins $16

YOW! These cartoons tell you how to make stuff. ZEEK! Kid stuff like marshmallow guns, water rockets, and duct-tape exoskeletons. POW! In serial comic form, you'll find explicit instructions on how to make, for instance, a working electric motor with just a C battery, a rubber band, two safety pins, and some wire. Or how to make ice cream in two plastic baggies. Neato! Kids who won't read "real" books will zip through these comic strips.

Despite the well-tested, foolproof instructions, young'uns will still need some adult assistance to complete the projects. At least our 11-year-old did. That's OK. This book will persuade them to do most of the work. And to try something more ambitious next time. After all, the projects not included in the book — the ones that kids think up themselves — are really the ones that this high-octane comic blast encourages most.

—*Kevin Kelly*

« Magical Machinery Tour

Recording The Beatles by Kevin Ryan and Brian Kehew
Curvebender Publishing $100

My hands were actually sweating and I was woozy with excitement as I unboxed my copy. Weighing in at 11lbs and 540 pages, this opulent hardcover tome (which comes in a faux studio tape box with oodles of extras, like photos, a lyric sheet, and other cool mementos) demands your attention. And any Beatles (or analog recording) fan will gladly sacrifice untold hours poring over its remarkable pages.

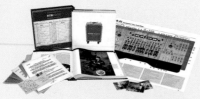

The authors (and self-publishers) spent a decade sifting through the notes and logs of EMI engineers and interviewing those involved in recording The Beatles. Every bit of technology used, from mics and instruments to recorders and mixing consoles, to the studio spaces themselves, is detailed in the text and shown in hundreds of photos and diagrams.

With the current resurgence of interest in analog recording and playback technology, connoisseurs will flip over all the photos of vintage equipment. Given the dizzying amounts of technical detail, you might think this book is only for recording engineers or engineer wannabes. Far from it: the authors never forget to tell the human stories, the creative side of things, and the lengths that the engineers went to in innovating and cobbling together new recording processes. The result is a publishing milestone, both a masterpiece of record engineering history and a significant piece of book-art worth every penny of its cover price.

—*Gareth Branwyn*

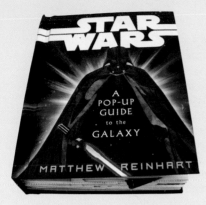

« Pop-Up Perfection

Star Wars: A Pop-Up Guide to the Galaxy by Matthew Reinhart
Orchard Books $33

When a preview copy of *Star Wars: A Pop-Up Guide to the Galaxy* arrived at our house, I can't even express how wickedly cool we all thought it was. Much like families of yesteryear gathered around the radio for their favorite program, ours gathered around the book to ooh and aah with abandon.

Even the Tweener loved it, and he doesn't think anything is cool. The 5-year-old asked to see it again for a bedtime story, and complained that it was too short once my wife was done opening the more than 35 pop-ups.

We're big fans of pop-up books around here, and this one is in a class all its own. The detail is simply stunning, with a level of intricacy I've never seen before. Think pop-ups on top of pop-ups, and when we opened the last page and Luke and Vader's light-sabers actually lit up, well, let's just call it a slam dunk for some geeky family entertainment. —Bruce Stewart

« Maker Math

Practical Math Application Guide by Norman Chenier
Chenier Enterprises $20

OK, so this may not seem like a great holiday gift, but trust me, it is. The math taught in this practical guide is the kind I wish they'd taught me in high school. Each section uses real-world situations (with a homebuilding bent) to illustrate different math principles, so you simultaneously learn trigonometry and how to lay out stairs, geometry and how to level with a plumb bob.

It has worksheets to practice what you've learned as well as "trade tricks" for DIYers that you wouldn't find in a math textbook. Many makers know this stuff intuitively already, but if you're new at this kind of thing, the book is a great resource. By the end, you'll know how to deal with unequal roof pitches, but you might just be a bit more of a math geek, too. —AO'R

» One-Stop Book Shop

Prices vary lulu.com

Lulu is a print-on-demand system that you can use to make, publish, and sell your own books. I used it to create a book for my nieces and nephew, and was really happy with the results. The process is pretty easy: download their templates, add your content, upload the templates, choose your book printing and cover options. Once you're finished, the books are available on demand, much like Café Press T-shirts. It takes a while for the books to print and ship, so if you need your books by a certain date, plan ahead! (*For more self-publishing options, see "Book Yourself," page 53*)

—Terrie Miller

Prototyping and projectiles — as chosen by the engineers of IDEO.

❮❮ McMaster-Carr mcmaster.com

When we have to build a prototype in a day and need that certain door latch, self-tapping screw, or compression fitting, we go directly to McMaster-Carr. You can get just about any industrial supply item, from micrometers, to furnaces, to O-rings, to 500-gallon storage tanks. And they offer overnight shipping on most things! This site is definitely in the category of "trusted resource."

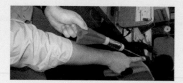

❮❮ Finger Blasters recfx.com

Need to blow off some steam? Ambush a colleague? Shock a client? We've found that Finger Blasters do the trick. Harmless foam projectiles (but be careful at close range), these finger-launched rockets are great for infusing the workday with a sense of play.

❮❮ FDM Rapid Prototyper stratasys.com

After staring at a computer terminal for days designing a part that's one-tenth of the size of what appears onscreen, there's nothing like actually holding the part in your hand to bring you back to reality. Our FDM (fused deposition modeling) machine gets nonstop use for that very reason. It is effectively a three-dimensional printer, taking the CAD-generated geometry created on the computer and building the part out of layers of ABS or other plastic in a matter of hours.

❮❮ Arduino Electronics Board store.makezine.com

Oftentimes we need to add electronics to our prototypes, but designing and building a custom-printed circuit board takes too much time and money. This low-cost board takes input from a variety of sensors, and can control LEDs, buzzers, motors, or anything else you care to connect to it. Best of all, you can order them directly from the Maker Store!

❮❮ Prototyping Team

One of IDEO's secret weapons is its team of master prototypers. If they can't build it, it can't be built. Admittedly, this option is a little too costly for your average DIYer, but these folks are definitely one of our favorite things. In lieu of your own private team of professionals, find a nearby shop with a variety of equipment and skilled machinists to help you out with that rare fabrication problem.

❮❮ Foamcore

This is the lifeblood of IDEO, along with Post-its. This sandwich of paper and foam is great not only for organizing our ubiquitous Post-its, but for creating quick, rough prototypes of everything from small, complex mechanisms to room-sized furniture installations. With foamcore, an X-Acto knife, and a hot melt glue gun, it's possible to build just about anything.

IDEO is an innovation and design firm that has been independently ranked as one of the most innovative companies in the world. Known for the design of Apple's first mouse, the first laptop computer, and the Palm V, IDEO more recently created the cockpit configuration for the Eclipse 500 Very Light Jet, Bank of America's Keep the Change banking service, and the Shimano Coasting bike strategy. ideo.com

My Sweet Ride

$50–$60 boardpusher.com

I got to try out boardpusher.com, where you design your own skateboard deck and they create it and send it to you. It was super cool. It's really great how you get to download images from your computer straight onto your board, to really make it your own. The choices of full-color backgrounds and designs were pretty good, and you can choose from six different board shapes, from old school to mini. You can also put your own message on the board (team name, maybe?), using one of 60 different fonts. BoardPusher is great for skaters of all ages. My dad said I almost peed in my pants with excitement when I found out I could get my own custom-made deck that could be whatever I wanted it to be — I'm not sure about that, but it is very cool!

—*Kindy Connally-Stewart, age 12*

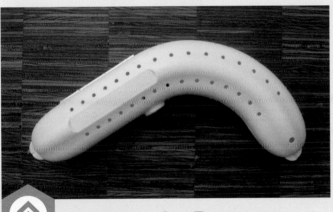

Bananas for Bananas

$7 bananaguard.com

Bananas, so nutritious and yummy, make for a perfect grab-and-go snack. But their elegant, naturally biodegradable and compostable wrappers cannot protect them from the ravages of my backpack. How many times had I scraped banana guts off the pages of a thick book or the vents of my laptop when I'd carelessly tossed them in with that vulnerable tropical fruit? Too many! That is, until I discovered the Banana Guard. Now, I no longer fear the big squish. Two of my friends had each mailed this clever carrier to one another at the same time, tickled by its, er-hem, suggestive shape. That shape can fit 90% of all bananas, and I've never had a problem sliding my packs o' potassium into its universal curve. Cubicle dwellers are advised to keep the BG in a bag or propped open to reveal its cargo, or else be ready to raise a few eyebrows.

—*Michelle Hlubinka*

Sky Calendar

$11 makezine.com/go/skycal

Sky Calendar comes as a one-page-per-month subscription from Abrams Planetarium at Michigan State University. The front of the page is a calendar that's packed with information about planets, moon phases, and other sky events during the month. The back has a map showing major constellations and planet locations during the evening throughout the month. When you subscribe, you get three months of Sky Calendar at a time. It's been around for years and continues to be a great resource for casual sky watchers and amateur astronomers alike. —*Terrie Miller*

Getting Junk Mail Stopped?

Priceless

Sometimes a real present is the kind that involves nothing at all. Spam is a pain, but it's gone with one click. My heart sinks as I see postal junk mail fill up my recycling bin day after day. And once you get on one mailing list, it's all over. But there are a number of sites that give you tips on reducing postal junk mail (try ecocycle.org/junkmail or obviously.com/junkmail), or you can give the gift that keeps on not giving, at greendimes.com.

—*AO'R*

Gareth Branwyn is a contributing editor at MAKE.

Kindy Connally-Stewart is a 12-year-old geek-in-training.

Unblemished fruit fuels **Michelle Hlubinka's** work as an illustrator, designer, educational consultant, and Maker Faire's community manager.

Kevin Kelly is the publisher of Cool Tools and is Senior Maverick for *Wired*.

Tim Lillis is a San Francisco-based illustrator, designer, musician, and aspiring professional zombie actor.

Bruce Stewart is a geek dad who blogs at geekdad.com.

Have you used something worth keeping in your toolbox? Let us know at toolbox@makezine.com.

From the Locksmithing Institute (1971)

■ **As any reader of old handy magazines** knows, the bulk of the advertisements were geared to individuals who wanted to increase their skills, wealth, independence, and perhaps most stirringly, the respect afforded them by their family. Make BIG money in plastic laminating, TV repair, lawnmower blade sharpening, tropical fish breeding, drafting, baby shoe bronzing, and cartooning! But did anybody ever make big money with home study courses? Did anyone ever make *any* money?

As I have a soft spot for amateur lockpicking, I jumped at a recent estate sale opportunity to buy the complete Locksmithing Institute home program. I quizzed the seller about her father Raymond's locksmith training. A mature and complex art, locksmithing is firmly in the 8th ring of handy. It's a magic skill only spoken of in reverential tones, and it greatly benefits from its delightfully closed society of practitioners and secrets.

In 1971, Raymond decided to become a professional locksmith to supplement his movie projectionist income. His daughter told me it took him three years to finish the coursework, due to his day job and duties maintaining his family's courtyard apartment building. Locksmithing became a valuable income stream, enabling Raymond to retire early and support his family on the proceeds of work completed in his meticulous home workshop.

Before I left, I visited his workshop. Estate sales demand that the house is disassembled and reconfigured to be a sort of antique store. The objects are still there but the context has usually been so disturbed that it offers little about how the owner lived. Workshops, however, are almost always left intact. The contents have less monetary value, and warrant only a cursory reorganization. As the workshop is not built to serve myriad family functions and impress the neighbors, it is often a deeply personal space. As I sat at Raymond's workbench, I was exceedingly proud of his efforts.

The locksmithing home study course is no trifle. Glad as I am to have the documentation, I would happily pay the $7.50 per course to get the materials, send my completed homework to HQ, and reap the benefits of a professional looking at my work.

All images compliments of Mister Jalopy's library

Mister Jalopy breaks the unbroken, repairs the irreparable, and explores the mechanical world at hooptyrides.com.

IMPORTANT
RETAIN AT KEY MACHINE FOR HANDY REFERENCE

INSTRUCT
Manu

HOW
TO
DUPLICA
A KEY

VOL. III
1956

Locksmith
Student Nu

		1st lesson due in, on or be $7.50 per month or lesson 32 lessons
$10.00		
1	7 50	
2	7 50	
3	7 50	
4	7 50	
5	7 50	
6	7 50	
7	7 50	(Dec 7 1971 pd in full)
8	0	
9	0	
10	0	
11	0	
12		
13		
14		
15		
16		
17		
18		

5 4 3

OIL CUP

CUTTER

RIGHT VISE

CARRIAGE

Fig. 9

...CKSMITHING INSTITUTE • LITTLE FALLS, NEW JERSEY 07424

Welcome To The World's Finest Home Study Locksmithing School!

Right now, while your first Lessons in Locksmithing are fresh in your hands, let's compare notes to make sure you receive all the benefits you are entitled to gain as a student in this practical and interesting School.

...to learn Locksmithing. You want to acquire the skills and know-
...oney at this trade. You want to build a solid reputa-
...smith.

...We are here to furnish
...in those skills and

LESSON CORRECTION

LOCKSMITHING INSTITUTE

Lesson No. 9

GRADE: "Complete"

The pin tumbler cylinder that you set
key is assembled well and operates correctl

Now that the "mystery" of the pin tumb
...has been revealed to you, I'm sure that you
...y key fitting job of this nature that come
...wever, I don't want you to become over conf
...takes practice to build experience.

It also takes practice to gain speed in t
...rk. I'd like to suggest that you visit y
...ard and try to purchase several old cyli
...you can experiment with them. This experi
...very helpful to you.

...eover, you will be building up a st
...s that you can use to make you
...ne more profitable. By hav
...you will be able to make
...In other words, if you ha
...ustomer's, you can give him yours and ta
...nge. This will give him quicker service
...nit you to fit the keys to his cylinder in
...time.

Walt Tiedeman
Chief Instructor

Lessons		Lesson mailed	Lesson due
Lessons	1	6-1-1971	6-5-1971
	2	7-2	7-3
	3	8-6	8-6
	4	9-3	9-10
	5	9-14	10-5
	6	9-28	11-5
	7	10-19	12-5
	8	3-29-1972	1-5-1972
	9	3-29	2-5
	10	4-12-	3-5
	11	4-14	4-5
	12		5
	13		6
	14		7
	15		8
	16		9
	17		10
	18		11

Project Orion: Saturn by 1970
By George Dyson

What if the 1950s had just kept on going?

■ **Fifty years ago tail fins, not seat belts,** were standard equipment on American cars. Russia was ahead in space, but America was ahead on the road. *Sputnik I*, weighing 184 pounds, was launched on Oct. 4, 1957, and circled the Earth every 90 minutes for the next three months. *Sputnik II*, weighing 1,120 pounds, followed on Nov. 3 and included Laika, the pioneer of spacefaring dogs. Earth's third artificial satellite was launched by a 32-ton Jupiter-C rocket built by the Chrysler Corporation, on Jan. 31, 1958. *Explorer I* weighed 31 pounds.

The race for space had begun. In Washington, D.C., the Advanced Research Projects Agency (now DARPA) was given a small office in the Pentagon and assigned the task of catching up. NASA would not exist until July of 1958. All three armed services had competing designs on space. "If it *flies*, that's our department," claimed the Air Force. "But they're called space*ships*," replied the Navy. "OK, but the Moon is high *ground*," answered the Army, who had already enlisted rocket pioneer Wernher von Braun.

ARPA's mission was to consider all alternatives, however far-fetched. One of the alternatives, code-named Project Orion, was an interplanetary spaceship powered by nuclear bombs. Orion was the offspring of an idea first proposed, as an unmanned vehicle, by Los Alamos mathematician Stanislaw Ulam shortly after the Trinity atomic bomb test at Alamogordo, N.M., on July 16, 1945. It was typical of Ulam to be thinking about using bombs to deliver missiles, while everyone else was thinking about using missiles to deliver bombs.

With Sputnik circling overhead, Los Alamos bomb designer Theodore B. Taylor (*see MAKE, Volume 07, page 188*), who had recently moved to the General Atomic Division of General Dynamics, decided Ulam's deserved another shot. "I was up all night and then I got alarmed that things were getting big," he recalls. "Energy divided by volume is giving pressure, so the pressures were out of sight, unless it was very big. It got easier as it got bigger."

According to Taylor, it was Charles Loomis, a former Los Alamos colleague working in the adjacent office, who said, "Well, think big! If it isn't big, it's the wrong concept." Taylor's perspective shifted. On Nov. 3, 1957, the day that *Sputnik II* (with Laika aboard) was

launched, General Atomic issued T.B. Taylor's *Note on the Possibility of Nuclear Propulsion of a Very Large Vehicle at Greater than Earth Escape Velocities*.

His proposal, submitted to ARPA in early 1958, envisioned a 4,000-ton vehicle, carrying 2,600 5-kiloton bombs. ARPA wrote a check for $999,750 to start things off. Suggested missions ranged from the ability to deliver "a hydrogen warhead so large that it would devastate a country one-third the size of the United States" to a grand tour of the solar system that Orion's chief scientists envisioned as an extension of Darwin's voyage of the *Beagle*: a four-year expedition to the moons of Saturn, including a two-year stay on Mars.

Could you blow something up without blowing it up? The answer appeared to be yes.

"Saturn by 1970," projected the physicists. "Whoever controls Orion will control the world," claimed General Thomas Power, commander in chief of the Strategic Air Command. An Air Force review of the project concluded: "The uses for Orion appear as limitless as space itself."

At the center of General Atomic's new 300-acre campus in La Jolla, Calif., near Torrey Pines, was a circular technical library, two stories high and 135 feet in diameter — exactly the diameter of the 4,000-ton Orion design. The library provided a sense of scale. Ted Taylor would point to a car or a delivery truck, the size of existing space vehicles, and say, "This is the one for looking through the keyhole." Then he would point to the library and say, "And this is the one for opening the door."

Orion was to be a one-cylinder external combustion engine: a single piston reciprocating within the combustion chamber of empty space. The ship, egg-shaped and the height of a 20-story building, is the piston, armored by a 1,000-ton pusher plate attached by shock-absorbing legs. The first 200 explosions, fired at half-second intervals with a total yield equivalent to some 100,000 tons of TNT, would lift the ship from sea level to 125,000 feet. Six hundred more explosions, gradually increasing

Images courtesy of James R. Burke, artist unknown (top); General Atomic (bottom left); and Thomas Macken, artist(s) unknown (bottom right)

Two Mars exploration vehicles in convoy, designed by General Atomic for NASA in 1963–1964; cutaway drawing of a U.S. Air Force military payload version of a 10-meter-diameter Orion vehicle; General Atomic and the Torrey Pines mesa in La Jolla, Calif., from the air, 1964, looking north.

in yield to 5 kilotons each, would loft the ship into a 300-mile orbit around the Earth.

"I used to have a lot of dreams about watching the flight, the vertical flight," remembers Taylor, who planned to be on board for the initial trip to Mars. "The first flight of that thing doing its full mission would be the most spectacular thing that humans had ever seen."

The 4,000-ton, single-stage Orion vehicle proposed in 1958 was intended to deliver 1,600 tons to a 300-mile orbit, 1,200 tons to a soft lunar landing, or 800 tons to a Mars orbit and return to a 300-mile orbit around the Earth. An "Advanced Interplanetary Ship," powered by 15-kiloton bombs, with a takeoff weight of 10,000 tons, was envisioned as 185 feet in diameter and 280 feet in height. Payload to a 300-mile orbit was 6,100 tons, to a soft lunar land-

ing 5,700 tons, or to a landing on an inner satellite of Saturn and return to a 300-mile Earth orbit — a three-year round-trip — 1,300 tons.

Nuclear fission releases a million times the energy of burning chemicals, but if you want to drive a spaceship, energy alone is not enough. Orion depended on whether translating the energy of a bomb to the momentum of a ship was feasible or not. Could you blow something up without blowing it up? The answer appeared to be yes.

Internal-combustion nuclear rockets are limited by the temperature at which the ship begins to melt. Orion escapes this restriction, because burning the fuel in discrete pulses and at a distance avoids high temperatures within the ship. In a chemical rocket, the burning fuel becomes the propellant. Orion's propellant can be almost any cheap, inert material

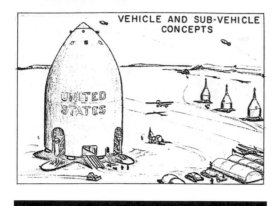

ABOVE: Sketch by Walter Mooney, 1963, of a Mars exploration mission. Note the lander/return vehicles and the bulldozer (which would be left on Mars). AT RIGHT: The 4,000-ton Orion vehicle, military payload version, 86 feet in diameter, ca. 1962.

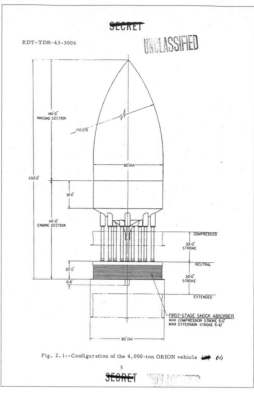

Fig. 2.1--Configuration of the 4,000-ton ORION vehicle

that's placed between the pusher and the bomb. It might be as light as polyethylene or as heavy as uranium, and, on a long voyage, might include shipboard waste in addition to ice, frozen methane, or other material obtained along the way.

The propellant is vaporized into a jet of plasma by the bomb. In contrast to a rocket, which pushes the propellant away from the ship, Orion pushes the ship away from the propellant — by ejecting slow-moving propellant, igniting the bomb, and then bouncing some of the resulting fast-moving propellant off the bottom of the ship. The bomb debris hits the pusher at roughly 100 times the speed of a rocket's exhaust, producing temperatures that no rocket nozzle could withstand.

For about 1/3,000th of a second, the plasma stagnates against the pusher plate at a temperature of about 120,000°F — hotter than the surface of the sun, but cooler than a bomb. The time is too short for heat to penetrate the pusher, so the ship is able to survive an extended series of pulses, the way someone can run barefoot across a bed of coals.

Even on an ambitious interplanetary mission involving several thousand explosions, the total plasma-pusher interaction time amounts to less than one second. The high temperatures are safely isolated, in both time and distance, from the ship.

For seven years, the Orion team worked on a series of increasingly detailed designs, but never received the green light to move forward to actual nuclear tests. "Over this entire time span, no technical reason has been found that would render the concept not practicable," they concluded in the project's final report.

It was impossible to separate the development of Orion from the development of bombs. In 1958, the U.S. was testing some 100 megatons of nuclear weapons in the atmosphere each year — and not getting any closer to Mars as a result. A full-fledged Orion mission would have added about 1% to the fallout from existing weapons tests. Orion's physicists believed this could be reduced by a factor of between 10 and 100 by designing cleaner — and highly directional — bombs. Still-classified evidence suggests they were at least partly right.

Orion remains the answer to a question we were never able to ask. Because of secrecy, the project was never opened to public discussion, and had no chance of gaining popular support. There was a narrow window of opportunity between the launch of *Sputnik* in October 1957 and the establishment of NASA in July 1958. As ARPA's role in space was brought to a conclusion, military missions went to the Air Force and peaceful missions to NASA.

Orion was orphaned as a result. The Air Force was reluctant to adopt a project aimed at peacefully exploring the solar system. NASA was reluctant to

Images courtesy of USAFSWC, History Office, Kirtland Air Force Base; and General Atomic

ABOVE: Tethered testing of a 300-lb 1-meter-diameter high-explosive driven flying model at the Point Loma test site, summer 1959. Present that day were Freeman Dyson (carrying briefcase) along with (clockwise from top) Ed Day, Walt England, Brian Dunne, Perry Ritter, Jim Morris, Michael Feeney, W.B. McKinney, Michael Ames. SEQUENCE AT RIGHT: Free flight testing, October 1959. The initial kick is given by a 1-lb charge of black powder; the subsequent 5 charges each contain 2.3 lbs of C-4.

adopt a project driven by bombs. Taylor's original vision — that Air Force officers would man the bridge while civilian scientists sampled the rings of Saturn — missed its chance.

"You will perhaps recognize the mixture of technical wisdom and political innocence with which we came to San Diego in 1958, as similar to the Los Alamos of 1943," Taylor's colleague Freeman Dyson, my father, wrote to J. Robert Oppenheimer on March 17, 1965, when the project finally came to an end. "You had to learn political wisdom by success, and we by failure."

In an Orion obituary notice that appeared in *Science*, my father, who had planned to be on board for the voyage to Mars and Saturn, elaborated: "What would have happened to us if the government had given full support to us in 1959, as it did to a similar bunch of amateurs in Los Alamos in 1943? Would we have achieved by now a cheap and rapid transportation system extending all over the Solar System? Or are we lucky to have our dreams intact?"

Adapted from the book Project Orion, *with new material. Part 2 of this article, "Project Orion: Deep Space Force" will appear in MAKE, Volume 13.*

George Dyson, a kayak designer and historian of technology, is the author of *Baidarka*, *Project Orion*, and *Darwin Among the Machines*.

Stills from 16mm film by Jaromir Astl

Euro Hacker Spaces
By Bre Pettis

Ich bin ein Hacker.

▪ I went on a tour of hacker spaces

across Europe this summer. Berlin is filled with hackers! C-Base (c-base.org), Chaos Computer Club (ccc.de), and Esch (eschschloraque.de) are all popular hangouts.

Starting at the CCC hacker camp near Berlin, I drove through the night to Austria just to go to the Metalab (metalab.at/wiki/English), an awesome hacker space in Vienna. It's located in the heart of the city, right near the parliament building. It's a vital and active space, and the Metalab folks have been intentional about setting it up and making things happen there.

They have facilities for presentations, community space for hardware and software hacking, a chill room for playing video games, a library, a foosball table, and a wet room for working on photography and circuit board etching. Paul Böhm, a founder, gave a great presentation describing the intentions, infrastructure, and development of their hacker space.

Located in the small town of Bochum, Germany, Das Labor (das-labor.org) is a hacker space filled with flashing LED gadgetry and vital hacking action. When the Hackers on a Plane group visited, Jörg Bornschein and Tilman Frosch explained that the laboratory was founded by accident. A group of cultural organizations decided to rent a building

▲ Bre's hacker tour of Europe started at the Chaos Communications Camp, where hackers from around the world gathered for presentations, food, camaraderie, and Club-Mate, the best soft drink ever.

and they needed one more tenant. Looking to Netzladen in Bonn (netzladen.org) and C4 in Cologne (koeln.ccc.de) for inspiration, they wrote up rules and brought together 20 people to discuss costs and membership fees. To spread the word, they postered local colleges. Now with approximately 42 members, Das Labor is a great space to work on electronics or software projects. Their work inspired a project for the Weekend Projects Podcast.

If this inspires you like it does us, check out our special online article on setting up your own hacker spaces: makezine.com/go/hacker.

Now back in the States, I'm starting up a hacker group called NYC Resistor, and we're on the lookout for a clubhouse where we can collaborate, learn, share, reverse engineer Club-Mate, and work on awesome projects (nycresistor.com).

Bre Pettis produces MAKE's Weekend Projects Podcast. Tune in every Friday afternoon and learn about a project you can make in a weekend at makezine.com/podcast.

Photograph by Bre Pettis

MAKE's favorite puzzles. (When you're ready to check your answers, visit makezine.com/12/aha.)

Poison Wine

An evil king has a cellar filled with 1,000 bottles of wine. An equally evil queen from a neighboring land plots to kill him and sends a servant to poison the wine. The king's guards catch the servant after he has poisoned 1 of the 1,000 bottles. The guards don't know which bottle was poisoned, but they know the poison is so strong that no matter how much it was diluted, ingesting any amount will be fatal. Furthermore, it takes exactly 1 month before it has an effect (after which it's fatal immediately).

The king decides he will use 10 prisoners to test the wine. Being an extremely clever king, how does he use his 10 prisoners to determine which bottle is poisoned, and still be able to drink the rest of the wine at his anniversary party in 5 weeks' time?

Black or White?

A Chinese emperor has to choose a new adviser amongst 3 sages, all of them equally wise and trustworthy. He tells them, "To choose one of you, you'll play a simple and fair game. In this sack there are 3 white balls and 2 black balls. Each of you will be blindfolded and will pick 1 ball and place it on your head. After that, the blindfolds will be removed and each sage in turn will try to guess the color of the ball upon his head by observation of the other picked balls. However, beware! You may pass your turn whenever you like, but once you state a color, if it is wrong you will fail and be disqualified. This way I'll learn which among you is the most intelligent."

The sages talk briefly to each other and promptly refuse. Because of their honesty, they know they cannot willingly mislead each other by providing false answers when they are not certain. "Emperor, it's of no use because the game is not fair and we must be honest with you and with each other. The third sage that guesses in the first round will always know the answer." The sages then promptly demonstrated this to the emperor, who was so amazed by their wits that he appointed all 3 as his advisers. How did they prove it to the emperor?

Michael Pryor is the co-founder and president of Fog Creek Software. He runs a technical interview site at techinterview.org.

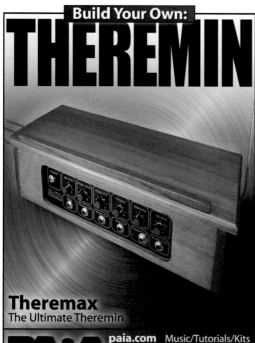

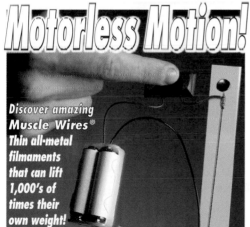

MAKER'S CALENDAR

Compiled by William Gurstelle

Our favorite events from around the world.

Jan	Feb	Mar
Apr	May	Jun
July	Aug	Sept
Oct	Nov	Dec

» NOVEMBER

☼ Roboexotica: Festival for Cocktail Robotics
Nov. 22–25
Vienna, Austria

Roboticists from around the world try to develop the latest and greatest in automated drink-mixing technologies. Event organizers describe it as "an index for the integration of technological achievements in everyday life, and ... a means of documenting the creation of new interfaces for man-machine interaction." It's also a cool way to get your martini made. roboexotica.org

» DECEMBER

» Shuttle Atlantis Rocket Launch
Dec. 6
Cape Canaveral, Fla.

Space shuttle *Atlantis* will blast off on the 24th U.S. mission to the International Space Station. The primary mission is to deliver the European Space Agency's Columbus Laboratory module.

makezine.com/go/atlantis

» JANUARY

» New Year's Whistle Blast
Jan. 1, Brooklyn, N.Y.

Steam aficionados from the Pratt Institute let loose during their annual steam-whistle extravaganza celebrating the arrival of the New Year. "Pulling the lever on the whistle from the *USS Normandy* and being enveloped in steam," say they, "is an experience not to be missed."

Also, listen to the blasts from railroad, riverboat, and really big ocean liner whistles. makezine.com/go/pratt
» Related project on page 88

» Delta 4 Heavy Rocket Launch
Jan. 25
Cape Canaveral, Fla.

The Delta 4 Heavy features three booster cores mounted together to form a gigantic, high-thrust, triple-body rocket. This mission will launch a classified spy satellite into orbit. makezine.com/go/delta4

» FEBRUARY

« Discover Engineering Family Day
Feb. 16, Washington, D.C.

This annual event attracts more than 6,000 visitors to the National Building Museum to take part in dozens of hands-on activities sponsored by national and local engineering organizations. eweekdcfamilyday.org

IMPORTANT: All times, dates, locations, and events are subject to change. Verify all information before making plans to attend.

Know an event that should be included? Send it to events@makezine.com. Sorry, it is not possible to list all submitted events in the magazine, but they will be listed online.

If you attend one of these events, please tell us about it at forums.makezine.com.

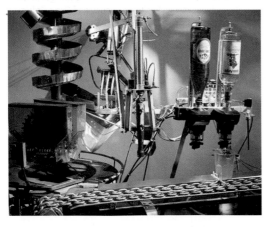

Photography by NASA (top two); F.T. Eyre, National Building Museum (middle); and Jacob Applebaum (bottom)

READER INPUT

Where makers tell their tales and offer praise, brickbats, and swell ideas.

I just wanted to comment on how much I love MAKE magazine and how much I appreciate all you guys do for us. The mag is absolutely fantastic and the podcasts are my absolute favorite. Keep up the great work; your content is some of the best on the entire net. —*Andrew Montgomery*

As a new subscriber, a novice maker, and a graduate student at UC Davis, I am impressed with the detail and care taken in publishing the project "Tabletop Biosphere" by Martin Brown (*Volume 10, page 110*). I have been working with exotic and invasive plant species to prevent and control their proliferation in both Florida and California. Let me express the appreciation of myself and my colleagues for your notation in the article reminding and educating your readers of the risks and proper disposal of any organisms they may obtain in the course of making. —*Thaddeus Hunt*

Editor's Note: **Thaddeus, thanks for the kind note. Our office biosphere was "sealed" on March 31, 2007. The ghost shrimp, we are happy to report, was alive and kicking as recently as Oct. 9! That's 6 months of biosphere goodness, and it gives us hope for successful larger-scaled biospheres in the near future.**

Finally got the paid subscription to MAKE magazine. First such purchase of a magazine subscription in 20+ years ... it's that good! —*Randy Poppe*

I have recently found your magazine ... it's great. And this is from someone who has never done any type of electronic project or task as I read in your magazine. Although I work in the computer industry, I am not an engineer. I love to tinker and will continue to read your magazine until I have the courage to make one of your projects. As a manager I was not able to get my hands dirty, as it were. Keep up the good work.

My suggestion, and something I would like but cannot build, is a RAM memory checker. The cheapest one out there is around $800. There are other problems besides the price: it gives too much information, for one. The two best features of a

MAKE LAB BIOSPHERE: Still thriving 6 months later!

RAM checker would be the ability to check all types — SIMMs, DIMMs, SDRAM, DDR, etc. — and the ability to Bluetooth to a printer. The simple check of the RAM would be sufficient: is the RAM OK?

Now, what I don't know is whether someone has already done this project. Any referrals would be welcome. —*Garrett Romain*

Editor's Note: **What do you say, readers? Got any ideas for Garrett? Post them at makezine.com/12/readerinput.**

MAKE AMENDS

The illustration of the ratchet assembly on page 118 in MAKE, Volume 11, is incorrect. Please refer to Figure 12 at makezine.com/11/birdfeeder to see exactly (and correctly) how it all goes together.

To summarize the error: The pawl does not go through the hole in the top of the pivot screw, and the ratchet spacers don't go over the pivot screw.

The battery in Figure A on page 152 in MAKE, Volume 11, was mistakenly identified as a lead-acid battery. It is, as many readers pointed out, a NiCd battery. We apologize for the mix-up.

In our MAKE: Halloween special edition, we attributed the photographs of HauntCon, pages 75 to 78, to Lillian Gurstelle. They should be credited to Karen Hansen.

On page 81 of the Halloween special edition, a 4-way 5-port valve (under the heading "Solenoid Control Valve, Body-ported") was identified incorrectly. We regret the error.

Sometimes it's not easy to decide which toys, tricks, or puzzles to include in these pages. Of course, any that are new to me seem appropriate, but a little research may reveal that they're not as novel as I thought. A puzzle I've never seen may be dismissed as "old hat" by puzzle collectors; I may even discover that it was commercially marketed. A physics demonstration may cause a teacher to comment, "That one was discussed in one of the physics journals, oh, maybe 15 years ago," with an implication that every teacher knows it, so it's not worth discussion.

Magic tricks are often marketed, but seldom patented. Anyone who is a bit curious can find out the secrets of small tricks and large stage illusions with a trip to the library. When you start digging, you discover that very few of these things are new, and many have precedents going quite far back in history.

So I am resisting dismissive reactions from insiders. What's common knowledge among experts is not necessarily known to the general public, and just might be of interest to quite a few MAKE readers; some may be inspired to make the devices I discuss here, or something like them, perhaps improving the idea in the process. The rest can skip this and move on to all the other good stuff in this magazine.

- -

An Elegant Puzzle

Many puzzles are based only on geometry or topology. A few require a principle of physics for their solution, and these have a special appeal for me. I especially like puzzles that have everything out in the open, with no hidden mechanisms. Naturally I like puzzles that people have constructed with their own hands.

Here's a beautiful puzzle (photos at right) shown to me by Mary Nienhuis, a retired high school mathematics teacher. It was made some years ago by Paul Hooker, then an industrial arts instructor at West Ottawa High School in Michigan.

The construction is of purpleheart and hard maple (it says so on the bottom), with a transparent lucite cover that allows you to see two marbles inside, separated by a wooden dowel. The marbles can move on curved ramps that have a depression partway up each slope. The object is to get both marbles to rest in those two depressions — simultaneously. Tilting or shaking the puzzle does no good. You can easily get one marble into its hole, but any attempt to get the other one in just dislodges the first one.

The Solution

Close examination reveals that the dowel in the center goes all the way through the maple piece and protrudes just a bit from the bottom. This provides a pivot so that the entire puzzle can be spun like a top, causing the marbles to rise up the ramps and settle into their respective depressions. The puzzle must be constructed so that it is very well balanced about the pivot point.

This homemade puzzle is the cleanest and most elegant application of this principle that I've seen. So far as I know, this particular design has not been marketed. Anyone who enjoys woodworking can easily make one, or use the idea in other designs.

Quite a few classic puzzles employ spinning as part of their solution. Marketed "centrifugal" puzzles of this sort include the original Spoophem, patented by Fred Swithenbank in 1913 and released by R. Journet & Company in 1929; Moses' Cradle by 1970s game maker Skor-Mor; and the S.S. Adams Co.'s Dipsy Ball. ThinkFun's All Uphill is still available in a plastic version for just a few dollars.

Donald Simanek is emeritus professor of physics at Lock Haven University of Pennsylvania. He writes about science, pseudoscience, and humor at www.lhup.edu/~dsimanek.

ABOVE AND AT RIGHT:
An elegant, handmade
pivot-action "centri-
fugal" puzzle. BELOW:
A few classic puzzles
that use spinning as
part of the solution.

Dollar Bill Greetings
By Tom Parker

Sometimes it takes more to buy it than to make it from the money itself.

↑ $4.33
Store-bought greeting card and envelope.

↑ $4.00
Dollar bill card and envelope.

Photograph by Sam Murphy

HOMEBREW

My Lego UAV
By Chris Anderson

I didn't have my children to justify playing with Legos, but it certainly didn't hurt. Yet after a while, we (OK, mostly I) wanted to do something with Lego that had never been done before. But what?

The answer came to me while out on a run, and it combined three geeky things that were on my mind: Lego, R/C airplanes, and gyro sensors. Suddenly it came to me. Gyros are what you use for autopilots. An R/C plane with an autopilot? That's an unmanned aerial vehicle (UAV), or drone. Bingo. We were going to build the world's first Lego UAV.

HiTechnic sent me a gyro sensor, and I got my first dose of reality. The sensors use tiny "rate gyros," which don't measure absolute position. To get them to actually keep a plane flying straight and level, you'd have to combine them with acceleration sensors and do a ton of gnarly math to get around inertial forces, drift, and other complications.

Then came the second flash of insight. Keeping a plane flying straight and true is a solved problem — companies such as FMA Direct sell "co-pilots" for around $100. So that just left navigation for the Lego. I found a plane (a Hobbico ElectriStar) that was big enough to hold the Lego Mindstorms controller, gears, sensors, the R/C system, and a Mindstorms motor geared to move an entire rudder servo back and forth.

When I started, there was no good way to read GPS data with Mindstorms. So I went with a proof of concept that used HiTechnic's compass sensor, and helped my then-9-year-old write a program that would just tell the plane to fly a square pattern. So far, so good. Fortunately there were several groups working on the Bluetooth GPS problem on Mindstorms. So we ported all the code from Mindstorms NXT-G to RobotC, which turned out to be pretty easy.

Today we have a fully functional Lego UAV. You give it GPS waypoints, take it off manually, then flick a switch on the R/C transmitter, and it flies to the coordinates you've entered. That's pretty awesome, but we want more. So the next job will be to integrate an onboard cellphone that communicates with the Mindstorms via Bluetooth. Cool, huh?

➕ Full story at makezine.com/12/homebrew.

Chris Anderson is the editor-in-chief of *Wired* magazine and runs a site on amateur UAVs at diydrones.com.

Photograph by Anne Anderson